THE HISTORY OF GLOBAL
CLIMATE GOVERNANCE

In 1979, global scientists concluded that climate change is a serious threat to humankind, and in 1992 the United Nations Framework Convention on Climate Change was adopted. However, more than 20 years later, although considerable progress has been made and the scientific community continues to provide evidence of climate change, we are still struggling to provide effective global governance to mitigate emissions and address its impacts.

What has happened globally on the climate change issue? How have countries' positions differed over time, and why? How are problems and politics developing on an increasingly globalized planet, and can we find a solution?

In this book, leading expert Joyeeta Gupta explores these questions and more, explaining the key underlying issues of the conflicts between international blocs. The negotiation history is systematically presented in five phases which demonstrate the evolution of decision making. The book discusses the coalitions, actors and potential role of the judiciary, as well as human rights issues in addressing the climate change problem. It explains that the process demonstrates progress but demands learning from past issues, and argues for a methodical solution through global law and constitutionalism, which could provide the quantum jump needed in addressing the problem of climate governance. This fascinating and accessible account will be a key resource for policymakers and NGOs, and also for researchers and graduate students in climate policy, geo-politics, climate change, environmental policy and law, and international relations.

JOYEETA GUPTA is Professor of Environment and Development in the Global South, in the Department of Geography, Planning and International Development Studies at the University of Amsterdam. She also works as part of the scientific steering committees of many different international programmes including the Global Water System project and the Earth System Governance project. Professor Gupta is editor-in-chief of *International Environmental Agreements: Politics, Law and Economics* and is on the editorial board of several journals, including *Carbon and Law Review, Environmental Science and Policy* and the new journal on *Current Opinion in Environmental Sustainability*. Her published work includes writing as a lead author for both the IPCC Report – which shared the 2007 Nobel Peace Prize with Al Gore – and the *Millennium Ecosystem Assessment*, which won the Zaved Second Prize, in addition to several books on climate change including *The Climate Change Convention and Developing Countries – From Conflict to Consensus?* (Kluwer Academic, 1997); and *Our Simmering Planet: What to do About Global Warming* (Zed Publishers, 2001). She is also the co-editor of *Mainstreaming Climate Change in Development Cooperation* (Cambridge University Press, 2010).

'Joyeeta Gupta's analytical history tour of a quarter-century of climate change negotiations provides a good entry point for newcomers and a pause for reflection by veterans. The former will benefit from the foundations of her analysis in climate science and sustainability, while the latter will be intrigued by her outreach to the domains of international law and human rights.'

MICHAEL ZAMMIT CUTAJAR, Executive Secretary, UNFCCC Secretariat, 1991–2002; Chairman of the Guardian of Future Generations, under the Sustainable Development Act (2012), Malta.

'This is an authoritative account of the history and complex political developments surrounding climate governance negotiation. Joyeeta Gupta has created a well-organized and interesting narrative, which will be extremely useful for anyone interested in climate change issues.'

DR YOUBA SOKONA, Board Member, South Centre; Coordinator of the African Climate Policy Centre (ACPC) at the UN Economic Commission for Africa (UNECA); co-chair of the IPCC Working Group III.

THE HISTORY OF GLOBAL CLIMATE GOVERNANCE

JOYEETA GUPTA

University of Amsterdam

CAMBRIDGE
UNIVERSITY PRESS

CAMBRIDGE
UNIVERSITY PRESS

University Printing House, Cambridge CB2 8BS, United Kingdom

Published in the United States of America by Cambridge University Press, New York

Cambridge University Press is part of the University of Cambridge.

It furthers the University's mission by disseminating knowledge in the pursuit of
education, learning and research at the highest international levels of excellence.

www.cambridge.org
Information on this title: www.cambridge.org/9781107040519

First published 2014

Printed in the United Kingdom by Clays, St Ives plc

A catalogue record for this publication is available from the British Library

Library of Congress Cataloguing in Publication data
Gupta, Joyeeta, 1964–
The history of global climate governance / Joyeeta Gupta, University of Amsterdam.
pages   cm
Includes bibliographical references and index.
ISBN 978-1-107-04051-9 (hardback)
1. Climatic changes–Law and legislation.   2. Climatic changes–Government
policy–History.   I. Title.
K3585.5.G87   2014
363.738'74561–dc23
2013028337

ISBN 978-1-107-04051-9 Hardback

To Hans van der Hoeven, without whose unconditional encouragement and support during the last 25 years my work on climate change and this book would have been impossible.

Contents

Foreword by Yvo de Boer   *page* xii
Preface and acknowledgements   xv
List of abbreviations   xvii

Part 1   Introduction   1

1   Grasping the essentials of the climate change problem   3
    1.1   Climate change intertwined with life   3
    1.2   Science, scientific uncertainty and climate sceptics   5
          1.2.1   The problem   5
          1.2.2   Sceptics and their rebuttal   5
    1.3   Climate change as an economic issue in an anarchic world   11
          1.3.1   A sceptical framing of climate change   11
          1.3.2   Countering this framework   13
    1.4   Climate change as a classic North–South issue   14
          1.4.1   The carbon budget and the ecospace problem   14
          1.4.2   GHG emissions and development   17
    1.5   Conclusion   21

2   Mitigation, adaptation and geo-engineering   22
    2.1   Introduction   22
    2.2   Climate change and development   22
    2.3   Options to deal with climate change   24
    2.4   Systemic changes   25
    2.5   Dealing with underlying driving forces   28
          2.5.1   Introduction   28
          2.5.2   Options and issues   28

2.6   Dealing with proximate driving forces: mitigation and
       sequestration options   29
       2.6.1   Introduction   29
       2.6.2   Options and issues   29
2.7   Dealing with atmospheric concentrations and warming   30
       2.7.1   Introduction   30
       2.7.2   Options and issues   30
2.8   Dealing with impacts   32
       2.8.1   Introduction   32
       2.8.2   Options and issues   32
2.9   Dealing with residual impacts   34
       2.9.1   Introduction   34
       2.9.2   Options and issues   35
2.10   Structure and outline of this book   37
2.11   Inferences   38

Part 2   The history of the negotiations   39

3   Setting the stage: defining the climate problem (until 1990)   41
     3.1   Introduction   41
     3.2   The chronology of events   41
     3.3   The problem definition and measures discussed   49
            3.3.1   The problem definition   49
            3.3.2   Measures discussed   50
     3.4   The role of actors   51
            3.4.1   Individual countries active: an agenda is born   51
            3.4.2   Science is institutionalized: an epistemic community
                    is born   53
            3.4.3   Other social actors   53
     3.5   The governance outputs   53
            3.5.1   The discourses: liability and the leadership
                    paradigm   53
            3.5.2   The principles   54
            3.5.3   The long-term objective   55
            3.5.4   Targets and timetables   56
            3.5.5   Policies and measures: winners and losers and no-regrets
                    measures   56
            3.5.6   Technology transfer – 'leapfrogging' and finance – 'new
                    and additional'   57
     3.6   Key trends in Phase 1   58

4 Institutionalizing key issues: the Framework Convention on Climate
   Change (1991–1996)   59
   4.1 Introduction   59
   4.2 The chronology of events   59
   4.3 The governance outputs   62
        4.3.1 Introduction   62
        4.3.2 The United Nations Framework Convention on
               Climate Change   62
        4.3.3 Assessing the Climate Convention   65
        4.3.4 The remaining INCs and the COPs   68
   4.4 The problem definition: leadership defined   69
   4.5 The role of actors   72
   4.6 Key trends in Phase 2   76

5 Progress despite challenges: towards the Kyoto Protocol and
   beyond (1997–2001)   78
   5.1 Introduction   78
   5.2 The chronology of events   78
   5.3 The governance outputs: the Kyoto Protocol and COP decisions   80
        5.3.1 The Kyoto Protocol   80
        5.3.2 Assessing the Protocol   85
        5.3.3 The COP decisions   91
   5.4 The evolving problem definition: conditional leadership   94
   5.5 The role of actors   95
   5.6 Key trends in Phase 3   97

6 The regime under challenge: leadership competition sets in (2001–2007) 99
   6.1 Introduction   99
   6.2 The chronology of events   99
   6.3 The governance outputs: the COP and CMP reports   101
        6.3.1 The Marrakesh Accords, COP7, 2001   101
        6.3.2 COP8 till COP10   107
        6.3.3 The COPs and CMPs (2005–2007)   109
   6.4 Climate-related agreements in other fora   114
   6.5 The climate funds over time   115
   6.6 The problem definition   118
   6.7 The role of actors   120
   6.8 Key trends in Phase 4   121

7 Enlarging the negotiating pie (2008–2012)   123
   7.1 Introduction   123

7.2	The chronology of events	123
7.3	The COPs and CMPs	124
	7.3.1 COP14–15, from Poznan to Copenhagen	124
	7.3.2 COP16–18, from Cancun to Doha	130
7.4	REDD revisited	135
7.5	Outside the regime	138
	7.5.1 Recession and climate change	138
	7.5.2 UN-REDD	139
	7.5.3 The human rights paradigm: countering market-based approaches	139
	7.5.4 Climate-related agreements in other fora	140
7.6	The problem definition	140
7.7	The role of actors	142
7.8	Key trends in Phase 5	143

Part 3   Issues in global climate governance   145

8	Countries, coalitions, other actors and negotiation challenges	147
8.1	Introduction	147
8.2	Countries and coalitions	147
	8.2.1 The formal classifications of countries	147
	8.2.2 Negotiation challenges	150
	8.2.3 The evolution of coalitions	151
8.3	Developed-country actors and coalitions	154
	8.3.1 The USA	154
	8.3.2 The European Union	156
	8.3.3 The East Bloc	159
	8.3.4 Russia	160
	8.3.5 Japan	160
	8.3.6 Others	161
8.4	Developing-country coalitions and countries	161
	8.4.1 The G77 and China	161
	8.4.2 Africa	162
	8.4.3 AOSIS	164
	8.4.4 LDCs	166
	8.4.5 OPEC countries	167
	8.4.6 China	167
	8.4.7 India	168
8.5	Other actors	169
	8.5.1 Non-state actors	169
	8.5.2 Sub-national authorities	170

8.6 Conclusions: the changing nature of the North–South discourse 171

9 Litigation and human rights 173
9.1 Introduction 173
9.2 The role of courts 174
9.2.1 Evolution: the rise of literature and legal action 174
9.2.2 The issues 177
9.3 Human rights and climate change 182
9.3.1 The evolution of human rights and climate change 182
9.3.2 The key issues 185
9.4 Inferences 189

Part 4 Towards the future 191

10 Climate governance: a steep learning curve! 193
10.1 Introduction 193
10.2 Framing 194
10.3 Learning 196
10.3.1 Unstructured problems call for learning 196
10.3.2 First-order learning: improving routines 197
10.3.3 Second-order learning: focusing on proximate drivers and impacts 199
10.3.4 Triple-order learning: focusing on underlying drivers and impacts 200
10.3.5 Implication: coherence where possible, leveraging elsewhere 202
10.4 Towards the rule of law at international level for climate governance 203
10.4.1 The rule of law as applicable to the project of climate governance 204
10.4.2 Global problems multiplying: scale of responsive governance increasing 206
10.4.3 Revisiting the fundaments of society: towards predictability 206
10.4.4 Limits of existing governance and incremental innovation 207
10.4.5 Pre-emptive geo-political pragmatism 209
10.4.6 Towards constitutionalization 209
10.5 Towards the future 210

References 212
Index 241

Foreword

YVO DE BOER

The climate change problem is a complex, multi-dimensional, systemic problem with a long temporal dimension. As a post-industrialization challenge, it has a long history; as its impacts will be felt long after the concentration of greenhouse gases have been stabilized, it has a long future. As it challenges our way of life, it is not just a technological problem, it is in essence a political problem. And as it ironically will affect all of us, it also has a way of bringing the whole world together. We are all dependent on each other for addressing the problem.

While in the past, the problem was essentially caused by the industrialized countries, this is now changing. The changing geo-politics of the world has brought new emitters into the picture. The victims of climate change remain the most vulnerable to the impacts of climate change. Although the problem is one that concerns the global commons, the way we share our rights to development with each other will remain the enduring problem, unless we can re-define and decarbonize our development pathways. This latter issue is not just a technological problem, it is deeply political – as it is politics that will create the space for economic investment in innovation by levelling the playing field. But levelling the playing field is not simply a matter of applying the same rules to all, it also means addressing global inequities.

This is what we have been negotiating about within the United Nations context. For the last 30 years we have given shape to the global discussions on climate change in the hope that we can find the way in which all our societies move towards a low-carbon future. The politics of negotiations shows how path-dependent this can be. Past behaviour shapes current intentions and current intentions will shape future behaviour. Any effort to understand the negotiating process will require an understanding of how past negotiations took place and how it shapes the current conflicts and opportunities. This is where this book attempts to shed light and provides an authoritative historical perspective of what has happened so far. Brought up in India, educated in the United States, and researching on North–South aspects, transatlantic aspects and the multi-level governance aspects of climate change both in the developed and the developing countries for the last 23 years in the Netherlands, Professor Gupta is well-placed to provide a cosmopolitan view of the negotiations as they have unfolded.

A key argument made in the book is the notion of the leadership paradigm. This was conceived early on in the pre-negotiation phase as a way to ensure that the developed countries would lead by reducing their own emissions and by helping developing countries to reduce their emissions. This was a positive paradigm as it focused on the technological and economic opportunities for change and not on finger pointing as to who was responsible and who should be held liable. It was conceived at a time when the developing countries were not yet committed to an irrevocable neo-liberal paradigm, their emissions were low, and their development pathways as yet undefined. It was conceived at a time when many technologies were available and only needed the political and economic space to develop. And yet the social contract underlying this paradigm was compromised by the inability of the United States to irrevocably commit to a stabilization target in 1992. This affected the political will in the rest of the developed world. By 1997, the United States helped the global community to design a regime that would mobilize the world, even if the United States itself was unable to participate effectively in the process. The leadership paradigm held on but was slowly beginning to crumble.

As this book argues, as time passes the room for emitting greenhouse gas emissions becomes smaller and smaller, if the global community wants to avoid dangerous climate change. This implies that the negotiating problem becomes more and more acute as there are less and less emissions to share between countries. This is particularly challenging when recession begins to hit the global economy – because while this causes a dent in the growth of emissions, it also reduces the commitment to address the problem as other issues such as youth unemployment and a struggling economy become priorities. Twenty years after the United States' unwillingness to accept a binding target in the Climate Convention, when Canada announced that it would also not participate in the post-2012 second Kyoto period, the domino effect was launched with Russia, Japan and New Zealand following suit. Leadership now rests uneasily on the shoulders of the European Union which is also struggling to cope with its own internal economic and financial crises.

Meanwhile, the world has changed. Some large developing countries have opened up their economies, embraced the neo-liberal path and are on the way to rapid growth. They are caught between feeling cheated by the failed leadership paradigm and the growing expectation that they should take action. They are caught between the feeling that they are 'damned' if they reduce their emissions, and 'damned' if they don't as the expected impacts are likely to hit hard. They cannot quite see their way out of this process. A vast majority of vulnerable peoples and countries are at the losing end of this tug-of-war in the negotiations on who should take what types of action first. This is where we are today.

And yet this book ends with a positive note. It argues that such a problem is unprecedented in human history and is a test of human ingenuity in being able to govern itself. It argues that it is inevitable that it will take time, because we are negotiating in a multi-speed world which includes extremely advanced countries, extremely vulnerable countries, and countries in conflict. It is therefore inevitable that the global learning process will take time. It argues that the mobilization of a global community may lead to the development of ideas and opportunities for addressing the problem as third-order learning sets it. It argues also that there is need for further developing legal tools – ranging from the recognition of human rights, to liability for causing harm to others, to global constitutionalization – as a way to bind humans together in this common endeavour. Law needs to provide a balance to politics to create the economic and technological space to solve a shared problem with huge equity implications.

Whether you agree with Joyeeta Gupta's interpretation of the history of climate governance or not, this book provides a short and compelling global overview of what has happened so far in the international negotiating process and the options for the future. It also shows how far we have come in first- and even second-order learning. The question is – are we ready for third-order learning?

Yvo de Boer is presently KPMG's Special Global Advisor, Climate Change and Sustainability. He was appointed by UN Secretary-General Kofi Annan as the Executive Secretary of the UNFCCC in 2006 and he held this position until 2010.

Preface and acknowledgements

I arrived in the Netherlands in 1988 and, upon completing a short assignment on the international trade in hazardous wastes, I was appointed by Pier Vellinga to help with national climate change policy and preparations for international negotiations. I was plunged into the preparations for the Noordwijk Conference on Climate Change and thereafter had to support the climate change department with its efforts at trying to develop both directional and instrumental leadership on the subject. Directional leadership was in terms of identifying the key principles, strategies and policies needed to help reduce the emissions of greenhouse gases and to show that it was possible and that it could generate co-benefits. Instrumental leadership was undertaken through a number of bilateral journeys to neighbouring countries, including those with economies in transition, to try and convince them to take on a stabilization target. Those were the heady days of supporting a proactive strategy in climate governance.

The speed with which the policy process moved in those days was unprecedented. Science was institutionalized and political processes were galvanized and, in a record two years, the United Nations Framework Convention on Climate Change was adopted. I moved out of the Environment Ministry to work on my PhD on climate change, and the interviews I conducted in those early days were very revealing. There appeared to be much conflict behind the consensus!

Over the 15 years since, the conflict has surfaced more prominently than ever. I remember that at Noordwijk, the US delegation had clarified that it shared the development aspirations of the developing countries. That has been the crux of the problem. All countries want to develop and few see a green, affordable way to do so. There are technological options available but the world is locked into an institutional and technological infrastructure. Developed countries do not wish to lose their competitive edge, developing countries want to catch up and if possible overtake. This underlies the dynamics of the negotiation process thus far. However,

I am very hopeful. Humans are ingenious and learning takes time. We will surely find the way to unlock the dilemma in which we find ourselves today. In the process, the vulnerable may increasingly become exposed to the impacts of climate change and it is necessary to find a way to both compensate them and to use them as a strong argument for mitigation action.

Having spent almost 25 years exposed to climate change, I have naturally made many friends along the way – and many of them did not hesitate to help with reviewing various sections of this book. I would like to take this opportunity to thank Karin Arts, Harro van Asselt, Laurens Bouwer, Catherine Brolmann, Nicholas Chan, Michiel van Drunen, Michael Faure, Antoinette Hildering, Hans van der Hoeven, Pedi Obani, Reyer Gerlagh, Eileen Harloff, Reggie Hernaus, Matthijs Hisschemöller, Onno Kuik, Mairon Bastos Lima, Henk Merkus, Leo Meyer, Arthur Mitzman, Hans de Moel, Charles Barclay Roger and the use of his database, Jacob Swager, Pier Vellinga and Wouter Wernink for their generosity and time in sharing their views and helping me to sharpen the analysis. Many thanks to my student assistants Marijn Faling and Andrea Brock. I think the biggest challenge has been what to leave out of the analysis – the history of the negotiations has been very rich and detailed. Ultimately I hope I have done justice to the story in the words that follow.

Abbreviations

AGBM	Ad Hoc Group on the Berlin Mandate (1995–1997)
AIJ	activities implemented jointly
ALA	Asian and Latin American (countries)
AOSIS	Association of Small Island States
AWG-DP	Ad Hoc Working Group on the Durban Platform for Enhanced Action
AWG-LCA	Ad Hoc Working Group on Long-term, Cooperative Action under the Climate Convention
AWG-KP	Ad Hoc Working Group on Further Commitments for Annex-I Parties under the Kyoto Protocol
BASIC	group of countries comprising Brazil, South Africa, India and China
CACAM	Central Asia, Caucasus and Moldova (group)
CB	capacity building
CBD	Convention on Biological Diversity
CBDR	common but differentiated responsibilities
CCAC	Climate and Clean Air Coalition to Reduce Short-Lived Climate Pollutants
CDM	clean development mechanism
CER	certified emission reduction
CGE	consultative group of experts
CMP/COPMOP	Conference of the Parties serving as meeting of the Parties
COP	Conference of the Parties
DC	developing country
EC	European Community
EIT	countries with economies in transition
ET	emissions trading

(EU) ETS	(European Union) Emissions Trading Scheme
EU	European Union
FCCC	Framework Convention on Climate Change
G7	Group of 7
G77	Group of 77
G8	Group of 8
GCOS	global climate observing system
GDP	gross domestic product
GEF	global environment facility
GHG	greenhouse gas
GNI	gross national income
IC	industrialized country
INC	Intergovernmental Negotiating Committee for the UNFCCC (1990–1995)
IPCC	Intergovernmental Panel on Climate Change
JI	joint implementation
JWG	Joint Working Group (SBSTA/IPCC)
KP	Kyoto Protocol
LDC	least developed country
LULUCF	land use, land-use change and forestry
MDGs	Millennium Development Goals
MRV	monitoring, reporting and verification
NAMA	nationally appropriate mitigation actions
NAPA	national adaptation programme of action
NGO	non-governmental organization
ODA	official development assistance
OECD	Organisation for Economic Co-operation and Development
OPEC	Organization of Petroleum Exporting Countries
QUELRO	Quantified Emission Limitation and Reduction Objective
REDD	Reducing Emissions from Deforestation and Forest Degradation (refers to RED, REDD, REDD+ and REDD++ in this book as an umbrella term)
RMU	removal unit (generated by LULUCF projects that absorb carbon dioxide)
SBI	subsidiary body for implementation
SBSTA	subsidiary body for scientific and technological advice
SIDS	Small Island Developing States
UN	United Nations
UNCED	United Nations Conference on Environment and Development

UNDP	United Nations Development Programme
UNEP	United Nations Environment Programme
UNESCO	United Nations Educational, Scientific and Cultural Organization
UNFCCC	United Nations Framework Convention on Climate Change
UNGA	United Nations General Assembly
USD	US dollar
WHC	World Heritage Convention

Part One

Introduction

1

Grasping the essentials of the climate change problem

1.1 Climate change intertwined with life

Ecosystems are interconnected (Commoner, 1971). Hydrological and bio-physical spheres flow into each other. Production and consumption patterns are interlinked across continents. 'National spaces, previously fragmented, are being integrated on a global scale' (UNIDO, 2008: 5). The ecological, social and economic crises are interlocking crises (WCED, 1987: 4) with intergenerational scope in an increasingly globalizing world!

Contrast that with our governance patterns. Local governance cannot cope with global externalities. National governance is affected by competing ideologies, interests and fragmented systems. Democratic politics is locked into 4–5-year recurring elections; while political decisions on institutions, technologies and infrastructure lead to long-term, locked-in, processes (Barbier, 2011: 238). Current investments are locking the world into an insecure, inefficient and high-carbon energy system (IEA, 2011). Lock-in refers to the difficulties in reversing decisions because of the high costs involved, and because these have a high inertia. Transboundary governance is affected by the competition between short-term national versus regional interests. Global governance is fractured along national interests, and is fragmented, pluralist, incoherent and often counter-productive. 'Glocal' (global to local) governance is affected by past politics. Path dependency affects the future.

Within these interconnected, interflowing, interlinked, integrated, interlocking and intergenerational crises is the climate change problem. The question that arises is, Should climate change be dealt with as a relatively small problem with clear contours or should it be addressed as a systemic problem? In 1989, Pier Vellinga of the Netherlands' Environment Ministry cautioned that the global community cannot afford to make climate change a systemic developmental problem because addressing the climate change problem would then become captive to addressing

3

all other global problems! Bert Metz (2010) cautions about politicizing the problem, arguing that a technocratic framing may yield better results. However, as the years have shown, climate change is far from a single-issue technocratic problem:

Furthermore, the suggestion that technical and normative considerations exist in separate universes is a fundamental misinterpretation of what climate change means as a driver of social change. It is not a 'natural' process that affects societies from the 'outside' – it is part of the metabolism of a socio-ecosystem. The challenge of building sustainable societies, in other words, cannot just be about technologies – or even institutions. The 'soft' infrastructures in the minds of members of society – their attitudes, beliefs and behavioural patterns – are intimately intertwined with the 'hard' infrastructures of steel and concrete through which we shape the world – and ourselves.

(Crowley, 2012: 3)

This chapter provides the context for the history of climate change governance. When the climate change problem was discussed in the 1980s, there were 'both North–South inequities and East–West tensions' (Toronto Declaration, 1988: Para. 14). The world has changed considerably since then. North–South inequities remain but the membership of the two groups has changed. East–West tensions have evolved (see Chapter 8).

A key message of this book is that the 'glocal' community is on a steep learning curve, it is moving from challenge to challenge – and this is promising! The problem is not technocratic – but very political: the way solutions are crafted or not crafted will have implications for 'who gets what, when, where, and how'! There is no avoiding the politics of climate change. However, political, social and technological solutions may well be in sight. Anil Agarwal of the Centre for Science and Environment in New Delhi once told me in a conversation that a global shift towards using renewable energy in place of fossil fuels would make the whole issue of sharing global resources, risks and responsibilities equitably ('ecospace sharing'; see 1.4.1) totally irrelevant! But in the meantime, we would have to focus on equity issues. With Brazil generating about 50% of its energy supply from renewables and Germany about 30%, we may be well on our way to making this revolution occur! Thus the emphasis on technocratic solutions is needed, not just in terms of adaptation and mitigation, but also in terms of re-engineering society in order to provide the trend-breaks society needs (Barrett, 2009). However, it will have to take place in the context of the politics of climate change.

Key to addressing a problem is understanding its nature (Hisschemöller, 1993). However, as problems are socially constructed, they cannot be defined 'objectively' and 'enduringly'. As this book demonstrates, problem definitions evolve as knowledge and perceptions develop. The science and its critique (see 1.2), the dominant framings (see 1.3) and the North–South realities (see 1.4) have all evolved.

1.2 Science, scientific uncertainty and climate sceptics

1.2.1 The problem

Anthropogenic or human-induced climate change is a post-industrialization problem caused by the net emissions of greenhouse gases (GHGs) such as carbon dioxide (CO_2), methane (CH_4), nitrous oxide (N_2O) and chlorofluorocarbons (CFCs) into the atmosphere. These gases emerge from the way we produce and consume. They emerge from our energy, agricultural, industrial and spatial planning systems. Water vapour and ozone in the troposphere and stratosphere are also GHGs. These gases envelope the earth and increase its temperature. The energy from the sun is the motor that drives the earth's climatic system. This energy arrives in the form of short-wave radiation of which about 70% is absorbed by the earth's surface and atmosphere. The earth also emits energy in the form of long-wave radiation. GHGs can absorb or re-radiate back this outgoing long-wave radiation as it is emitted from the earth, warming our planet further. Overall, the earth is about 30° warmer because of this GHG effect. By adding additional GHGs to the atmosphere, the accumulated concentration of these gases may lead to the *enhanced* global warming effect. This warming may change global climate patterns. However, there are a number of other elements that can also reinforce or negate the warming effect (see 1.2.2).

This global warming leads to expansion of the waters in the seas (imagine a boiling kettle), melting glaciers, changing wind and rainfall patterns, salt water intrusion into coastal areas as the sea level rises, and possibly extreme weather events. Beyond a certain point it can lead to non-linear irreversible changes – also referred to as 'tipping points' or crossing 'planetary boundaries' – the melting of polar ice and the slow-down of ocean circulation patterns. For a more nuanced and detailed analysis see the reports of the Intergovernmental Panel on Climate Change (IPCC). The IPCC assesses the work of scholars, building on the initial premises of Joseph Fourier in 1824 who postulated that there was a greenhouse effect; John Tyndall in 1861 who analysed the role of water vapour; Svante Arrhenius who argued that a CO_2 doubling in the atmosphere could lead to a few degrees of warming; and Guy Callendar who, in 1938, argued that CO_2 concentrations were indeed increasing in the atmosphere.

1.2.2 Sceptics and their rebuttal

Climate change is a global-scale inadvertent experiment. Although the basic relationship between increased concentrations of GHGs and warming is undisputed, it is not always clear how sensitive the earth's climate is to such concentration build-up. This is reflected in the language of uncertainty. Natural scientists can try to evaluate this uncertainty 'objectively' while social scientists argue that uncertainty is a social construct – i.e. an idea that society defines (Jassanoff, 1990; Shackley and Skodvin, 1995). This uncertainty can be used to justify action through the precautionary principle – i.e. the argument that even if there are doubts about the links between cause and effect, if the effect may eventually be irreversible, this irreversibility justifies action to minimize the cause. This is a dominant argument that frames the European Union's (EU's) perspective on climate change. At the same time, this uncertainty can be used to justify inaction or postpone action by those who argue that the costs of current measures to deal with the cause are too high, and possibly in the future these costs will come down. This is an argument used by the US government. However, as far back as in 1989, it was argued by countries participating in the Tata Conference Statement (1989: Art. 5.5) that 'If nations delay actions in an elusive quest for scientific certainty, the risks and costs will mount unacceptably'.

Part of the political problem of making decisions is the rise of climate sceptics. In the pre-1990 period there were scarcely any sceptics. However, by 1996 the sceptics were organizing themselves, arguing that climate models did not adequately take into account the impacts of water vapour and other feedback effects, that the models did not reflect the reality of the global system and that IPCC reports reflected political and not scientific consensus (Emsley, ed., 1996). There has been a gradual intensity in the rise of scepticism in the post-2000 period. With 1998 being the warmest year in recorded history up to that point, many began to play on public ignorance by arguing that the scholars were manipulating the data and were deliberating making mistakes to prove their own hypothesis. This scepticism has been predominantly present in the USA, and some say that this is politically motivated (Bowen, 2008; Mooney, 2006). Similar attacks have been launched to question the integrity of the scientists participating in IPCC, their email exchanges and some mistakes in the IPCC reports, but subsequent reviews of IPCC work have shown that their basic conclusions are correct.

The main arguments of the climate sceptics and their rebuttals can be clustered as follows (see Table 1.1). First, the contribution of anthropogenic emissions is marginal compared with natural causes. Anthropogenic emissions are a mere 3–4% of total emissions of GHGs; and events such as solar variation, volcanic eruptions and the El Niño Southern Oscillation and other natural variability can cause greater changes and related problems. Anthropogenic emissions are also not significant in terms of global time-scales: through history the climate of the earth has been changing, the temperature has fluctuated and there have been ice ages.

However, most natural causes are in balance with the effects; many natural causes (e.g. solar variation, volcanic eruptions) cannot be controlled while human causes can be controlled; and the issue is not whether the earth will have a problem

Table 1.1 *The arguments of the sceptics and their rebuttal*

Arguments	Rebuttal
Anthropogenic emissions marginal	Small but significant
Only 3–4% of total GHG emissions, natural emissions more important	But natural system can cope with natural emissions; not the anthropogenic increment
Not significant in geological time-scales	But we worry about now and our children
Solar cycle variation can influence by 0.2–0.4°	Solar variation small in comparison with expected climate impacts
Volcanoes, El Niño, meteorite hits are important	These cannot be controlled
Impact of anthropogenic emissions marginal	Unprecedented and has non-linear impacts
Radiative effect small – CO_2 doubling leads to 1° rise; 2° rise is no big deal	Disregards feedback mechanisms, rise is unprecedented in the last 10,000 years, regional variations high, non-linear impacts; the temperature increase is already at 0.8°C and there is more in the pipeline because of delayed feed-back effects
Evidence of warming unconvincing	Scientifically accurate
Upper atmosphere is colder	But lower atmosphere is warmer
Measurements are in warmer urban areas	But are corrected for heat island effect
Decades of CO_2 emissions did not lead to warming (1940–1960; since 1998 the world has been getting cooler; 2008–2010: cold winters in US/Europe)	1940–1960: because of cooling effect of sulphates; post-1998 averages are higher than pre-1998 averages; very warm in Greenland, big temperature variations linked to complex feedback processes
There have been ice ages in the past	Linked to changes in the tilt of the earth's axis; next ice age expected in 20,000 years
Model generalizations cannot capture reality	The models are getting better over time
IPCC work – political consensus, not fact	IPCC work – scientific convergence
Warming is not necessarily a problem	Long-term winners are unpredictable
Warming is good: enhanced precipitation, longer growing season, increased plant growth, melting of Arctic opens transport/mining options	Regional variations problematic for some; non-linear impacts will be problematic for all
If a problem, can be dealt with by adaptation and geo-engineering	There are limits to adaptation, and geo-engineering has many side effects
Mitigation measures are problematic	Depends on how they are designed
Ineffective: Sea level will continue to rise for centuries; positive feedback effects	Hence, need for early action

Table 1.1 (*cont.*)

Arguments	Rebuttal
Expensive: Is too expensive, leads to leakage and loss of competitiveness	Not necessarily, depends on design of response system
Disruptive: World economy will collapse	Not necessarily
Diverts scarce resources from global priorities	Climate change impacts on global priorities,
It is cheaper to take action later – new technologies	But the problem set in motion may be irreversible
The science is self-serving	The critique is also self-serving
Helps climate scientists, actors and big government supporters remain in power, in line with doomsday thinking	Helps neo-liberals, small government supporters, technology optimists and GHG-producing industry retain power

in geological time-scales but whether humans are creating a problem for current and following generations. Moreover, with a changing baseline, the effect of natural variability may become increasingly more difficult to handle causing, for instance, extreme conditions that were previously very rare.

Second, the impact of anthropogenic emissions is miniscule and the signature of anthropogenic emissions against the background noise is difficult to detect. The radiative effect of CO_2 is limited; a CO_2 doubling in the atmosphere in relation to pre-industrial levels leads to a maximum of 1°C rise in temperature (Rahmstorf, 2009: 38). This may appear easy to deal with for the public – take off a sweater; adjust the thermostat!

However, this is without considering any feedback mechanism in the climate system (like the ice-albedo effect – where warming leads to melting of ice and thus decreases the albedo or reflection of the heat by the ice and leads to more warming). By how much temperatures would increase exactly with a CO_2 doubling from pre-industrial levels is subject to scientific debate, but about 3° seems reasonable though some argue that we are moving to a 4° rise. Furthermore, such a rise is unprecedented over the last 10,000 years, occurs over a very short time span, and a mean rise in temperature hides huge spatial variations. These spatial variations may exacerbate the situation of vulnerable lands and peoples, and can lead to non-linear irreversible impacts such as the melting of the Greenland ice sheet, boreal forest die-back in the USA and Russia, instability of the West Antarctic Ice Sheet and changes in the Indian monsoons (Lenton *et al.*, 2008).

Third, sceptics argue that the evidence of warming is not convincing. For example, there is a time-lag between emissions of GHGs and the resultant warming of about 5 months (Kuo *et al.*, 1990). This is not consistent with historic time-lags through ice-core studies (Petit *et al.*, 1999: 433). Furthermore, the temperature is increasing in the lower atmosphere while it is much colder in the upper atmosphere (Schwartzkopf and Ramaswamy, 2008). Moreover, the causal relationship is problematic as there have been very warm periods on earth in the past without a corresponding increase in anthropogenic GHGs; there have recently been decades that were relatively cool (e.g. 1940–1960), and temperatures have not reached the record of 1998 since then. Model generalizations simplify reality into a caricature and the IPCC consensus reflects political consensus, not scientific fact. The predictions of doom of the Club of Rome (Meadows, 1972) did not materialize.

However, carbon isotope analysis shows that current concentrations are from fossil sources; and there are explanations for aberrations in warming related to climatic variations linked to, for instance, the solar cycle and the El Niño Southern Oscillation. For example, between 1940 and 1960 there was less warming probably because of the cooling effect of sulphates emitted by thermal power plants into the atmosphere (Mitchell and Johns, 1997, Mitchell *et al.*, 2001). The year 1998 was very warm because of natural variation due to a very strong El Niño in this case; that this record has not been topped yet does not mean that the earth is cooling. While accepting that laboratory models do not represent reality, it is impossible to set up a global scale experiment; past predictions are only inaccurate when they have led to policy measures to avoid the outcome predicted; and the past predictions of the Club of Rome (Meadows, 1972) with respect to CO_2 emissions and climate change have more or less come true (Vellinga, 2012: 29). The real uncertainty lies in knowledge regarding the sensitivity of the climate system to GHG emissions. The climate system could have moderate sensitivity if there are more sinks that absorb GHGs than we know of, if a CO_2 doubling is not eventually accompanied by increased water vapour, and if increased cloud cover leads to cooling. The ice ages are caused by changes in the tilt of the earth's axis, and the next ice age is not expected for 20,000 years.

Fourth, the accumulation of GHGs is not necessarily problematic – as this can increase precipitation in some regions, lengthen the growing season, encourage plant growth through enhanced CO_2 concentrations, and thereby be beneficial. For example, carbon fertilization could lead to a net benefit of USD 37 billion to the USA (Mendelsohn *et al.*, 1994). History shows that during the period AD 850–1350, there was enhanced warming which led, on balance, to more food, trade and better health. The melting of the Arctic opens up options for transportation and mining. There are, for those who hope to be winners, no reasons to frame the climate change issue as a 'problem'. If the problem turns out to be serious, one can always adapt or apply geo-engineering methods.

However, it is difficult to downscale global changes to local levels – the climate may become unpredictable; in the long term the non-linear effects are definitely problematic for society and may threaten human life on earth; geo-engineering addresses primarily the symptoms of climate change and has many side effects (see 2.7).

Fifth, the planned measures are ineffective, expensive, disruptive and not a priority. The targets adopted in the Kyoto Protocol of 1997 (see Chapter 5) contribute to barely effecting a 0.1°change in temperatures over a hundred-year period. Sea-level rise will continue long after measures are taken. The measures proposed are expensive, may lead to carbon leakage to other parts of the world and will affect the competitiveness of industry. Significant measures to deal with climate change can disrupt the economies of the world without having any additional benefits. None of this implies that countries should not invest in measures that can have other benefits. The money can be better spent on other issues such as addressing developmental challenges (Lomborg, 2001)! Measures could perhaps better be postponed to a time when the new technologies are cheaper (Wigley *et al.*, 1996).

However, measures need not be ineffective, expensive and disruptive or come at the cost of development priorities. This all depends on how the measures are designed (see Chapter 2).

Finally, promoting climate change as a problem is seen as self-serving for specific groups of actors – it helps to generate resources and power for climate scientists and climate actors, and it appeals to some neo-Marxists, supporters of big government, and those who see themselves as 'losing' from the climate problem.

However, this argument is equally self-serving for neo-liberal, small government supporters, and GHG-intensive industry and consumers who are afraid to 'lose' if climate mitigation is emphasized (see Table 1.1). For more details see Vellinga (2012).

I believe that there is enough evidence that climate change is a serious problem (IPCC reports; Joint Statement of Academy of Sciences, 2001; NRC, 2010), and that the uncertainty in the science does not necessarily disprove the causal links (van der Sluijs, 1997; Vellinga, 2012). However, the media apparently need to present debates rather than facts in order to generate discussion and viewership, thereby often providing a platform for two opposing views even when the views may not be equally authoritative. The media often undermine the authority and legitimacy of the IPCC, recognized through the award of the Nobel Peace Prize in 2007, in the search for 'balance' with the views of other stakeholders. This search for 'balance' creates a bias (Boykoff and Boykoff, 2004) and confuses the public. This public is also becoming increasingly sceptical of modern scientists, seeing them as the witches of the modern era, as Professor Peter Haas has been noted to state. A sudden cool event is seen as evidence that climate change is a hypothesis, not a fact.

The bottom line is that in such a scientifically complex problem, it is important to have anchoring devices that help to maintain consensus amongst scientific scholars even if the knowledge base evolves (van der Sluijs, 1997). There may be a need to develop a best practice for constructing serviceable truths (Guston, 2001) which provides the basis for action in line with the precautionary principle.

1.3 Climate change as an economic issue in an anarchic world

1.3.1 A sceptical framing of climate change

If climate change is so serious, then why are governments reluctant to take action? On 27 January 2012, an article in *The Wall Street Journal* argued that there was 'No Need to Panic about Global Warming' – as the science is incorrect and the financial impacts low. In general, arguments against climate action are that (a) the macro-economic costs of stabilizing GHG concentrations at 350–500 ppm could be as high as trillions of USD (IPCC-III, 2001, ch. 8; Manne and Richels, 1997); (b) the benefits of mitigation have a limited impact over the next 40–50 years; (c) the costs of climate impacts (about 1.5–3.5% of GDP; IPCC, 1997) are a tiny proportion of the costs of restructuring the total economy; (d) including a better evaluation of the damage costs to life and ecosystem capital and services would still lead to a calculation that these costs are relatively small (although this is often more an implicit argument brought up in discussions); (e) the costs of 'climate migration' (which may involve 15–200 million people) are limited compared with the global autonomous population growth of a few billion; and (f) including even occasional punitive or compensatory damages to other countries as a result of possible court cases will not change the picture very much. (Besides, the impacts are exacerbated by inappropriate development and adaptation strategies and other factors such as shifting of the tectonic plates.) The reduced costs also emerge through the use of the market rate of discount (rate of return on capital), which reduces the current value of future impacts. The market rate of interest will not adapt to take climate change into account. Furthermore, people don't care for climate change enough to want to change their behaviour; they are not 'willing to pay'. Their choices are reflected in the market.

Moreover, countries with legally binding targets under the climate treaties may have to adopt stringent technology and production standards, leaving space for those without similar targets to use older technologies and production standards that are cheaper and more competitive in the international market. This might lead to a shift in the source of the emissions from one group of countries to another – the issue of leakage, which could in turn affect the competitiveness of developed-country economies and their employment opportunities (Byrd–Hagel Resolution, 1997, see 5.2). A focus on global net benefits and cost effectiveness just does not justify action. Hence, from a purely economic perspective, there is no 'rational' justification for taking *far-reaching* action, especially if such action aims at completely restructuring the economy into a sustainable development economy. This may destabilize national economies and possibly even the global economy. In fact, more knowledge about the benefits and costs of climate mitigation may reduce the incentive of parties to effectively participate in global climate governance (Kolstad, 2005; Kolstad and Ulph, 2008), although financial transfers to other parties may enhance participation (Dellink and Finus, 2009; Finus and Pintassilgo, 2011). Furthermore, the enduring (social) media conflict between climate believers and sceptics may reduce the motivation of policymakers and society to act (Dannenberg *et al.*, 2011).

A dominant political science framing is that the global system is anarchic and sovereign states may or may not participate in global cooperative ventures based on national interests. There is no exogenous power of constitutional norms and enforcement mechanisms that can force action (see 10.4). States cooperate only if the benefits of cooperation are greater than those of non-cooperation. However, where differentiated cooperation strategies are chosen, this means that some states may have to do more than others. Those who have to take more action see this as unfair, as other countries 'free-ride' or 'benefit' from their action without mitigating the problem themselves. The concept of 'free-riding' is especially important in relation to 'open access' goods – goods that can be used by all countries/actors because they are not 'owned' (Hardin, 1968; Stavins, 2011), leading to overuse of the resource (e.g. grazing lands, fishing waters) or the 'tragedy of the commons'. This results in a preference for either non-cooperation strategies or smaller fora, which allows for greater flexibility and efficiency (but also for externalizing issues of concern to other groups – leading to non-decisions, à la Bachrach and Baratz, 1970). Hence many argue that a multilateral approach is just not likely to work (Newman *et al.*, eds., 2006).

Some ecological modernization theorists also argue in favour of technocratic and market-based approaches. They focus on the need for sustainable households, industrial transformation, corporate social responsibility and social movements as a way to engineer change in society (cf. Mol and Sonnenfeld, 2000; Olsthoorn and Wieczorek, 2006). What they tend to have in common with mainstream economics is the tacit understanding that power structures should remain untouched – i.e. they don't analyse the structural societal causes of problems.

This combined framing of the climate change issue calling instead for bottom-up approaches challenges centralized, far-reaching global action.

1.3.2 Countering this framework

The above framing can be countered. First, mitigation costs may seem enormous at trillions of USD, but the reduction in income per year is minor in comparison with the expected rise in global income (Azar and Schneider, 2002; 2003) and it is cheaper to reduce emissions now than later (Nordhaus, 2008). Second, damage costs are probably underestimated through the use of standard formulae (e.g. the damage function) and the market discount rate (Azar, 1998), monetizing the cost of human life, undervaluing the damage to ecosystems, assuming that compensatory and punitive damages are unlikely to arise, ignoring the cost of loss of nationhood of small island states and citizenship (Limon, 2009), and the difficulty of valuing the cost of low-probability, high-risk events (cf. Azar and Schneider, 2003). Choosing a market rate of discount is inappropriate for assessing climate impacts (Azar, 1998; Nordhaus, 2008; Stern, 2006; Vellinga, 2012). Third, the influence of the linear and non-linear impacts of climate change (including the security impacts as assessed by national security agencies) on the production factors in society is ignored. Fourth, for neo-classical economics it doesn't matter who wins or loses – as long as the winners can compensate the losers. However, where winners and losers may live in different countries and regions, this may implicitly put the risks on those who live in low-lying areas or cyclone-prone areas, and compensation may simply not cover their losses (Azar and Schneider, 2003). Furthermore, it is more than possible that all countries will suffer damages at higher temperature increases (Stern, 2006). Fifth, carbon leakage is a possibility especially in a few energy-intensive sectors (e.g. cement, steel, aluminium), but it can be tackled through screening of energy-intensive sectors, border tax adjustments that are compatible with World Trade Organization rules, and international sectoral or global agreements. In fact, Van den Bergh (2010) lists 12 reasons why a climate policy is affordable!

Although the global society lives in an 'anarchic' world, global public goods (goods that are non-rival – where one person's use does not diminish the ability of another to use it, and non-exclusive – where no one can be excluded from its use, such as a stable climate) cannot be dealt with by the market (Kaul *et al.*, 2003). These need to be addressed through a system of governance. Management of the 'commons' is possible at local (Ostrom, 2011) through to global levels (Weiner, 2009), although it may be difficult. Issue-based negotiations with differing power constellations may provide room for negotiation (Junne, 1992), lower transaction costs may incentivize participation (Keohane, 1984) and cooperative processes can institutionalize norms, rules and rule-compliant behaviour (Krasner, 1982). A well-designed institution has an inertia and countries continue to behave in accordance with the process, and thus such regimes can have staying power. This explains that in a global society, countries are motivated, and obligated by international law, to cooperate. However, such cooperation may require structural leadership (reflecting economic, political and issue-related power), directional leadership (reflecting ideas, technologies and actual implementation) and instrumental leadership (reflecting the ability to make coalitions and link issues together) (Grubb and Gupta, eds., 2000; Malnes, 1995; Underdal, 1994; Young, 1991). As this book shows, the nature of leadership has changed over time.

Such leadership will have to go beyond the economics of climate change to allow understanding of the social justice aspects. As Sinden (2007: 271) puts it:

> Thinking of climate change as a human rights issue can help us see that it is not just a matter of aggregate costs and benefits but of winners and losers – of the powerful few preventing the political system from acting to protect the powerless many. But perhaps even more importantly, treating climate change as a human rights issue simply begins to imbue it with a sense of gravity and moral urgency that communicates to all of us: this is something different; this is an issue that must be understood to stand apart from the normal clatter and noise of day-to-day politics and to demand attention from our best selves.

In the meantime, the human rights agenda is once more being invoked by social justice movements and politicians. '[C]limate change raises profound questions of justice and equity: between generations, between the developing and developed worlds; between rich and poor within each country. The challenge is to find an equitable distribution of responsibilities and rights' (Miliband, speech, 2006) (see Chapter 9).

1.4 Climate change as a classic North–South issue

1.4.1 The carbon budget and the ecospace problem

If society wishes to ensure that climate change does not become 'dangerous' to humans, then GHG concentrations need to be stabilized in the atmosphere. Stabilizing these *concentrations* implies significantly reducing GHG *emissions* over time; this is needed as many GHGs have a long life in the atmosphere (see Figure 1.1). Furthermore, the amount of annual permissible emissions will keep decreasing and the sharing problem will thus become more intense over time.

IPCC-I (1990) stated that if atmospheric GHG concentrations were to be stabilized at 1990 levels (when the C concentration was 350 ppmv), then global emissions of CO_2 would have to be reduced by 60%, the emissions of methane by 15–20%, and nitrous oxide and CFCs by about 70–85%. IPCC-I (1990: 22) argued that 'Stabilisation at any of the concentration levels studied (350 to 750 ppmv) is only possible if emissions are eventually reduced to well below 1990 levels'. The IPCC Synthesis Report (IPCC, 2007: 66) added that:

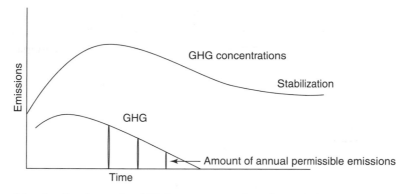

Figure 1.1   Implications of stabilizing concentrations for emissions

Stabilization at 445–490 ppm CO_2 equivalent concentration (including GHGs and aerosols) means that global emissions should reach their peak year between 2000 and 2015 and emissions of CO_2 must be reduced by 50–85% of 2000 emission levels. In this scenario, global temperature will rise by about 2–2.4°C above pre-industrial levels and the sea level may rise by between 0.4–1.4 metres. Stabilization at 490–535 ppm means that global emissions must reach their peak between 2000 and 2020; in this scenario temperature increases may be as high as 2.4–2.8° C and sea-level could rise by between 0.5 and 1.7 metres.

This implies that the permissible emissions in every year are limited. This is the carbon budget that needs to be shared between countries and peoples (see Figure 1.2). If society wishes to stabilize concentrations at a particular level by a particular year, then this gives rise to a 'safe landing' corridor of action. This implies that every year, all things being equal, the size of this budget decreases (see Figure 1.1).

This raises questions: First, how big should this budget be and how should the discussions regarding the size of the budget evolve? This is the question of the long-term goal of climate policy. When does climate change become dangerous to society and who defines when it has become dangerous to whom, and whose definition counts? The size of the budget can also be increased by relying more on adaptation and geo-engineering (see Chapter 2).

Second, the budget size is linked to development (see 2.2). In general, the more developed an economy, the greater the slice of the budget used and claimed by countries. Where countries aim to grow, they claim larger slices of the budget. Conversely, small economies use smaller slices of the budget; however, they would like to claim more of the budget! This brings us to the overarching issue of whether societies should be allowed to develop, and under what conditions (see 1.4.2).

Third, can we postpone the emission reduction process, hoping that new technologies will make this process cheaper in the future (Wigley *et al.*, 1996: 240–3)? However:

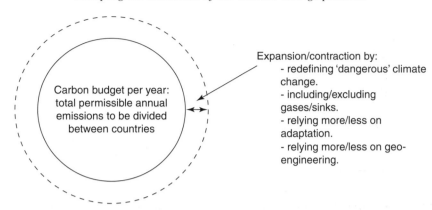

Carbon budget per year:
total permissible annual
emissions to be divided
between countries

Expansion/contraction by:
- redefining 'dangerous' climate
  change.
- including/excluding
  gases/sinks.
- relying more/less on
  adaptation.
- relying more/less on geo-
  engineering.

Figure 1.2   The carbon budget

delayed emission reductions significantly constrain the opportunities to achieve lower stabilisation levels and increase the risk of more severe climate change impacts. Even though
benefits of mitigation measures in terms of avoided climate change would take several
decades to materialise, mitigation actions begun in the short-term would avoid locking
in both long-lived carbon intensive infrastructure and development pathways, reduce the
rate of climate change and reduce the adaptation needs associated with higher levels of
warming.

(IPCC Synthesis Report, IPCC (2007: 6))

Fourth, how can this budget be shared between countries, sectors, peoples, over
years and over generations? Figure 1.3 shows how northern and southern emissions may grow in a business-as-usual scenario (left graph) and the questions
regarding how their future emissions should be curtailed under a policy scenario
(right graph). In determining how these emissions are to be curtailed, the critical
issues are: What principles can be used to share this budget? How is development
accounted for in these principles? Is the underlying logic of these principles losing
its validity over time?

Fifth, sharing this budget is challenging because of how global trade influences
the process. The trade in products and services which embody carbon and other
GHGs (i.e. products whose production process is carbon intensive and located in
one country and which are exported to other countries) could change the dynamics of how the budget is to be divided between countries. This becomes even more
complicated if one wishes to account for the production emissions associated with
the consumption of goods that take place elsewhere in the world.

Sixth, is it theoretically possible for countries to reduce their GHG emissions?
A shift in society from an industrial to a renewable energy-based service economy could help to reduce emissions through dematerialization and decarbonization of products, processes, logistics and services. But it is not automatic (see
1.4.2).

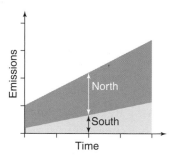
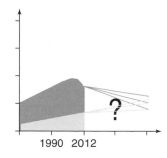

Figure 1.3   The long-term conflict

This is referred to broadly as sustainable development and/or the green economy. However, while it may be easier for the developed countries to achieve dematerialization, many developing countries who use older technologies may see a materialization taking place before the dematerialization sets in. At the same time, there is considerable potential for decentralizing the energy supply system and making more use of locally available renewable forms of energy using local capital and labour. Such a focus may also help to decarbonize developing countries.

Seventh, If we cannot decide on the size of the budget or how to share it, then adaptation becomes increasingly more important. This leads to issues such as who will have to adapt, who can adapt, who does not have the resources to adapt, and how should responsibilities for adaptation be dealt with.

Eighth, dividing a budget leads to someone winning and someone losing. This is distributive bargaining. The question is: Can integrative bargaining, which expands on the issues being negotiated, help to address the vexed issue of 'how to share the budget'?

Finally, can the postponement of emission reduction and the troublesome issue of adaptation be addressed through geo-engineering (see 2.7)?

1.4.2 GHG emissions and development

Patterns of development: Is sustainable development automatic?

Since climate change is linked to development (see 2.2), the critical issue is: how do societies develop? Demographic transition theories (Thompson, 1929; van der Tak *et al.*, 1979) posit that a transition from a rapid population growth stage to stable population is possible via intermediate stages of falling death rates and birth rates. Development theory (Rostow, 1960) submits that societies evolve from subsistence economies through specialization, industrialization and trade surplus, towards becoming service societies. Forest transition theory (Mather, 1992; Rudel *et al.*, 2005) explains that societies initially maintain their forests, then begin deforesting in order to develop (forest frontier stage), reach a forest and agricultural mosaic stage before they return to appreciating the role of forests and stabilizing their forests. These transition theories depict past patterns. There is no automatism in these pathways – e.g. countries get stuck in poverty traps: 'Economic development has been a historical exception' (Sachs, 2004: 10), populations continue to grow and growth does not by itself lead to stable forests (Box 2.1 in Gupta *et al.*, 2013). Such patterns are influenced by context-specific driving forces and policies.

On the environment, the Environmental Kuznets curve (Grossman and Krueger, 1995; see also Jänicke *et al.*, 1989; Malenbaüm, 1978 for material use; Grossman, 1995; and Selden and Song, 1996 for pollutants) submits that environmental pollution will increase as income per capita increases; after a critical point, income per capita will increase but will not be accompanied by that specific kind of pollution because society will be willing and able to spend to reduce the pollution. However, subsequent research shows that dematerialization may be followed by re-materialization (De Bruyn, 1998) and that material flows in economies may be growing (Berkhout, 1998) – possibly with a growth in demand. In fact there is evidence that climate change emissions grow with economic growth and there is no automatic incentive to reduce these emissions (Caviglia-Harris *et al.*, 2009; Copeland and Taylor, 2004; De Bruyn and Opschoor, 1997; Dinda, 2004; UNDP, 2011). Figure 1.4 integrates these pathway theories.

With increased wealth, societies invest in reducing indoor air pollution and increasing access to water and sanitation services. Up to a point, increased wealth leads to deforestation and higher urban pollution; beyond a critical level in income, forests are stabilized and urban pollution is reduced. However, although there are occasional bouts of delinking GHGs from economic growth, the volume of societal demand increases, leading to more GHGs (see Figure 1.4). This pattern will continue until society introduces specific policies.

How to develop?

The following question is: How do societies develop? The understanding of what promotes development has evolved over time (Easterly, 2007; Meier, 2001; Thorbecke, 2006). It has moved from a single-minded focus in the 1950s on increasing GDP by industrial and infrastructural investments (Nurkse, 1953; Rosenstein-Rodan, 1943; Rostow, 1956) to balanced growth, intersectoral linkages and regional integration in the 1960s. The impacts on equity and employment led in the 1970s to a focus on real per capita GDP and the Human Development Index (HDI). The role of privatization and liberalization became more important in the 1980s, followed by emphasis on poverty reduction, entitlements and capabilities and enhancing freedom (Sen, 1999), on the one hand, and sustainable development (WCED, 1987; see 2.4.1), on the other. But there are no clear formulae, as the current recession demonstrates.

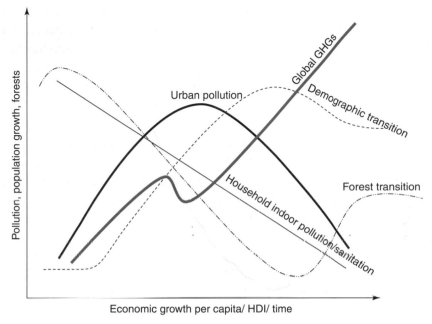

Figure 1.4   Theories of how societies develop. HDI, Human Development Index

The right to develop

If there is confusion about how to develop sustainably, this confusion is a breeding ground for discontent between developing and developed countries.

Just as economic growth is not automatic (Sachs, 2004) so also the road to sustainable development will not be an autonomous one but will require conscious steering. Although it sounds simple, the issue of who may develop is politically sensitive in the post-colonial world. While the UN Charter (UN, 1945) aims at promoting development in all countries, many newly independent societies experienced the stress of depleted resources, mines in foreign hands and the loss of social capital and governance infrastructure. They were struggling to rise up from the ashes of colonial life. They negotiated in favour of a new international economic order (Schrijver, 1995), the right to develop (Garcia-Amador, 1990) and economic and social human rights (Kirchmeier, 2006). The new international economic order aimed at countering the perceived existing North–South inequities in the global political and economic order (trade, debt, investment), but its legal instruments remained unimplemented (Schrijver, 1995). The right to development discussed since 1977 matured into a formal UN General Assembly Declaration on the Right to Development (UNGA, 1986), and is continually discussed in the UN. This right could be used by developing countries to justify a larger share of the carbon budget (Gupta, 2010b). This is increasingly contested especially within the climate negotiations.

Climate change as a classic North–South issue

Climate change is a classic rich–poor issue. The rich produce and consume more and their emissions are accordingly higher, and they are better able to cope with the impacts. The poor produce and consume less, have lower emissions and are more vulnerable to climate change, and are on 'the front line' (UNDP, 2007: 27). And the poor are often the factories of the rich – their emissions are related to consumption in rich countries. The consequences of increased emissions and its impacts such as extreme weather events damages every society worldwide – Katrina and Sandy demonstrated this in the USA. The issue of how the carbon budget is to be shared between sectors and households, and how adaptation resources are spent in different parts of any country at some future date, will expose this problem even further.

This issue is true not only in the national context but also in the supranational context. The issue of 'burden or effort sharing' in the European Union (EU) refers to how the responsibility for reducing emissions should be allocated between the EU member states – some rich and some relatively poor. This has been a 20-year debate within the EU and the original solidarity is beginning to crack under the burden of the financial crisis. The challenges of the EU may reveal the kinds of problems that may be faced when regions begin to allocate targets amongst themselves.

At a higher level of aggregation, climate change is the quintessential North–South problem, adding to scars from colonial days. The North is shorthand for rich industrialized societies, South for poor developing countries. This framing was included in the Climate Convention of 1992 (see Chapter 4). However, there were no criteria to define these groupings and this has become an enduring problem (see 8.2).

Nevertheless, climate change is a classic North–South problem as the bulk (70%) of present and past emissions since 1850 have come primarily from the developed countries. Even though the emission levels in China and some other developing countries are increasing rapidly, the impacts of these will be felt much further into the future. Besides, the per capita emissions of China and India were 1/5 to 1/15 of the emissions of the USA (UNDP, 2007: 7), but these figures are changing over time.

Furthermore, the bulk of present and near-future impacts will be felt primarily by the poor in the developing countries (IPCC, 2007; UNDP, 2007: 41), exacerbating their current vulnerabilities (Tata Conference Statement, 1989). Moreover, if the climate change problem is to be addressed satisfactorily in the near future, then UNDP (2007: 7) estimates that the global budget of emissions will be exhausted by 2032, leaving little room for the emissions of developing countries to grow.

However, having said that, the global system is navigating towards a problematic future. The developing countries can no longer afford to wait until the developed countries take far-reaching action even if it affects their sense of what is fair. For even if developed-country emissions return to zero, developing-country emissions are likely to be above 'tolerable' limits (Metz, 2010: 80). The right to develop is more a principle of fairness than an idea that is practical if the climate change problem is to be addressed.

1.5 Conclusion

This chapter has tried to set the tone of the book. It has argued that addressing climate change may call for restructuring society. The question then is: Is the problem real enough? This chapter has presented the arguments of the believers and the deniers – submitting that the climate science is strong enough to withstand the bulk of the arguments of the sceptics, and that the media with their need for viewer statistics often enlarge the space for sceptics at the cost of the science. It has argued further that the coming together of neo-classical economists, neo-realist international relations scholars and the ecological modernization movement has led to an emphasis on the need for industry to internalize costs and exercise corporate social responsibility in the hope that such actions will together address the problem. The counter-movement argues that this is not enough – a quantum leap in governance is needed to cope with the market imperfections that accompany public goods problems. Finally, the chapter argues that although what is North and what is South is changing, climate change is a quintessentially rich–poor problem. This rich–poor dimension is likely to play out at global through to local governance levels.

2

Mitigation, adaptation and geo-engineering

2.1 Introduction

The previous chapter presented the three big political dimensions of the climate change problem – the politics of science, the politics of governance and the politics of cooperation between rich and poor. This chapter now looks at the options for dealing with the climate change problem. It first argues that climate change is a development problem and that mitigation, adaptation and geo-engineering have impacts on development (see 2.2). It then looks at the measures that can be taken at different points of the climate change chain (see 2.3). For example, it examines measures to deal with the systemic nature of the problem (see 2.4), the underlying driving forces (see 2.5), the sources of climate change (see 2.6), GHG concentrations and global warming (see 2.7), adaptation (see 2.8) and residual impacts (see 2.9).

2.2 Climate change and development

Climate change is intrinsically linked to development. In order to address climate change, one can adopt mitigation measures to reduce GHG emissions and enhance sinks (e.g. afforest) that absorb these gases, adaptation measures that deal with the impacts of climate change, or adopt geo-engineering measures (large-scale interventions in the climate system) (see 2.7). Climate change mitigation and development have five key links. First, most societies and peoples wish to develop and the economic growth rate and employment are key to most election promises. Enhanced production, distribution and consumption brings with it increased GHG emissions. Second, emancipating the poor and providing them with access to basic resources, as called for in the Millennium Development Goals (UNGA, 2000a), will enable their meaningful participation in social production and consumption processes. This can potentially lead to enhanced GHG emissions, although it improves the resilience of the poor to climate impacts. Acceptance of the right of the poor to emancipate themselves implies acceptance of the need to differentiate between survival and luxury emissions (Agarwal and Narain, 1991; Opschoor, 2009). Third, development itself can lead to the conditions that can allow for mitigation: the rate of growth of these emissions can be influenced by the sustained application of appropriate policy choices, the use of win-win measures and the interconnectedness with other related policy decisions (IPCC-III, 2007). However, there is a limit to incrementally changing patterns of production and consumption in a society: structural change may be necessary (Biermann *et al.*, 2012a). Fourth, such mitigation measures can have ancillary benefits for development. For example, reducing fossil fuel use reduces GHG emissions. It also reduces the loss of foreign exchange for energy importers with soft currency. Reduced fossil fuel use has the co-benefit of reducing other related pollutants that may contribute to air pollution. However, new techniques have given an impetus for accessing oil and shale gas resources such that North America may become an exporter of such resources by 2030 (WEO, 2012). Fifth, measures to enhance sinks may lead to a decline in deforestation and forest degradation rates, which may improve the ecosystem services these provide to local societies and thereby enhance the opportunities for development.

Thus climate change mitigation is intimately linked to most sectors (e.g. energy supply, transport, buildings, industry, agriculture, forests and waste) in society and is thereby linked to infrastructure and settlements, employment, national income and investment, and lifestyles and culture.

Adaptation is 'adjustment in natural or *human systems* in response to actual or expected climatic stimuli or their effects, which moderates harm or exploits beneficial opportunities' (IPCC-II, 2007: 869). Adaptation is linked to development in three ways. First, measures to emancipate the poor may enhance their social capital, their resilience and their ability to cope with change (IPCC-II, 2007: 813). Second, measures to enhance development (mal-development) may themselves lead to increased vulnerability to the impacts of climate change (Tearfund, 2006). Third, adaptation measures can lead to benefits for development. Climate change will have impacts on other sectors: energy (e.g. via the impact on water flows), agriculture (via the impact on land use), industry (via the impact on water, food and other inputs), forestry (e.g. through changing (micro-)climate and increasing risk of forest fires), tourism (e.g. via impacts on beaches), health (e.g. by influencing the spread of disease) and transport (via e.g. the impacts on ports). Through such impacts, there will be impacts on income, welfare, employment, infrastructure and investment, culture and lifestyles, and life itself. The vulnerability to such impacts is influenced by local capacities – related to gender, age, wealth, social strata, etc.

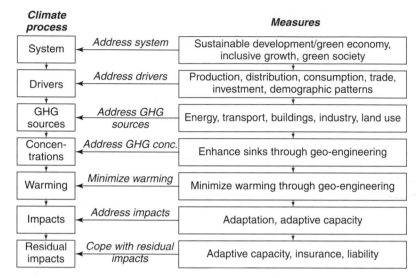

Figure 2.1   Measures targeted at different parts of the climate system

Geo-engineering and development are also closely related. More developed societies are better able to explore such opportunities. However, geo-engineering is also being explored in the Chinese context in relation to seeding clouds (Li *et al.*, 2013). Therefore, these options may be exclusively used to benefit these more developed societies. Nevertheless, the negative effects of geo-engineering may impact other countries as well (see 2.7).

2.3 Options to deal with climate change

In order to deal with climate change, it is useful to understand the climate change process chain (see Figure 2.1). Society functions within a political, economic, social, cultural system. Within this system, a number of underlying and proximate drivers (sources) influence the emissions of GHGs. The concentration of these GHGs in the atmosphere is the main factor influencing warming. The warming leads to a series of impacts (that can be dealt with) and residual impacts in different parts of the world.

Societies can invest in systemic change by promoting sustainable development, the green economy, inclusive growth and the green society (see Figure 2.2, and 2.4). This is also referred to in the literature as mainstreaming climate change in development. Societies can also 'unpackage the problem' and try and deal with individual issues. The underlying drivers of climate change are production, distribution, consumption, trade and demographic patterns. Addressing these calls for a series of specific measures (see 2.5). The proximate drivers or sources of climate

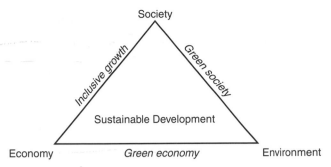

Figure 2.2   Sustainable development and its relation with green economy, green society and inclusive growth.
Source: Reproduced with permission from Gupta *et al.* (2013), Advisory Council on International Affairs

change are probably the easiest to deal with. This calls for identifying a series of measures to deal with the energy, transport, industry, building, and land-use sectors (see 2.6). One can also address the problem of GHG concentrations in the atmosphere through GHG removal directly from sources or from the atmosphere. GHGs help to absorb heat; preventing the solar rays from reaching the earth or reflecting the heat back through geo-engineering methods may be an option here (see 2.7). Finally, there are impacts. Societies then have to invest in adaptation measures and adaptive capacity (see 2.8). The residual impacts can be dealt with through insurance and liability (see 2.9). The sections below elaborate on these options in brief.

2.4 Systemic changes

The environmental impacts of development processes have been discussed since the 1960s. To deal with these impacts, the idea of a sustainable society (Brown, 1981) and the notion of sustainable development were conceived (WCED, 1987). Sustainable development is progress 'that meets the needs of the present without compromising the ability of future generations to meet their own needs' (WCED, 1987: 43):

> In essence, sustainable development is a process of change in which the exploitation of resources, the direction of investments, the orientation of technological development, and institutional change are all in harmony and enhance both current and future potential to meet human needs and aspirations.
>
> *(WCED, 1987: 46)*

Sustainable development requires a political system that allows citizen participation, an economic system that is self-reliant and sustained, a social system without tensions; a production, distribution and consumption system in line with the resource base; a sustainable international trade and finance system; and an administrative system that allows for self-correction (WCED, 1987: 65). The concept is included in the Rio Declaration on Environment and Development (1992), the Climate Change Convention (UNFCCC, 1992), the Convention on Biological Diversity (CBD, 1992) and Agenda 21 (1992), all adopted at the 1992 UN Conference on Environment and Development.

The concept implies the protection of resources for future generations while still meeting the needs of current generations and simultaneously addressing social, economic and environmental goals (see 1.4.3). However, although it is vague, like democracy and legitimacy, it can be equally inspiring (Lafferty, 1996). Sustainable development is not just an academic concept, it has strong North–South connotations – is it about promoting northern interests at the cost of southern development; or about protecting southern interests at the cost of the northern future? This has often led developing countries to reject the concept of sustainable development as being only achievable after a certain level of development has been reached, in line with the Environmental Kuznets curve (see 1.4.2). Sustainable development is an end and a means (Dovers and Handmer, 1993; Mebratu, 1998; Sachs, 1999) and promotes human well-being (Dasgupta, 1993; Sen, 1999). Strong sustainability implies meeting economic, environmental and social criteria while weak sustainability embodies trade-offs between these criteria and often ignores the social and political aspects (Banuri *et al.*, 2001; Barnett, 2001; Lehtonen, 2004; Robinson, 2004). This has led to the call for a social justice approach (see Chapter 9) as a way to strengthen the social criteria. If environmental, ecological and social issues can be effectively integrated, society will be on the path towards sustainable development. However, we do not know what this path looks like and so the story is about 'navigating through an uncharted and evolving landscape' (IPCC-III, 2007: 693).

Progress with respect to sustainable development has been limited. There is now talk of the green economy (UN, 2012; UNEP, 2010) which focuses on two of the three pillars – environment and the economy (see Figure 2.2). The green economy is seen as necessary, efficient and affordable (World Bank, 2012). It focuses on clean energy, modern technologies and reducing waste. It is promoted by UN agencies (UNEP, 2010) and is seen as multi-interpretable by the UN Secretary General (2010: para. 44), ranging from: (a) internalizing environmental externalities; (b) a systemic incorporation of environmental challenges within the economic order; (c) linking social with economic goals; to (d) moving towards a new macro-economic framework that designs a pathway towards sustainable development. For some, achieving a green economy is possible through tweaking the system by ensuring that prices reflect environmental externalities; by 'greening' infrastructure; taking the life cycle of a product into account in its design; dematerialization and decarbonization; developing sustainable spatial planning options; sustainable agriculture; protection of ecosystem services; promoting sustainable state procurement policies and ecolabelling to influence consumer demand; empowering and educating people; ensuring a rights-based approach to access to basic resources; and making states responsible for protecting the environment (EC, 2011; Nationaal Platform Rio+20, 2011; UN Secretary General, 2010). A key way to deal with both sustainable development and poverty alleviation may be through decentralized renewable energy systems.

For others, such tweaking of the system will not be enough to prevent the large-scale ecological degradation it causes (Barbier, 2011), and to cope with the new resource scarcities. The bottom line is that the green economy focuses on growth (Brand, 2012). It ignores social issues and the need for resilience building (Maxwell, 2012). The competing concept of inclusive development is now being promoted by the G77 and is also included in the Rio+20 document. It focuses more on social inclusion, development-led employment strategies and wealth for society. Although not clearly defined, it appears to mean 'growth coupled with equal opportunities' and focuses on growth, a level political playing field, a safety net, and rising inequalities (Rauniyar and Kanbur, 2010a: 455). The term development (rather than growth) emphasizes welfare rather than just an increase in per capita income, and could perhaps be reflected in the Human Development Index concept (which includes health and education). Inclusiveness refers to improvements in poverty reduction, and in reducing income differentials; it has a specific focus on income inequality (Rauniyar and Kanbur, 2010b) and focuses on the poorest in society (Sachs, 2004: 15). This calls for the exercise of rights, access to welfare programmes and access to civic amenities.

For many, since the work of the Club of Rome, this focus on growth is problematic. The concept of green society refers to a society that lives within ecosystem resources, is in harmony with nature, focuses on welfare and is not directed at economic growth per se. Some argue that continuous global growth comes with a growing 'ecological footprint', which measures the demands made on natural capital, to society. Post-modernists see it as an 'ideological trap' that perpetuates inequalities between and within states (cf. Sachs, 2004: 9). Green society responds to the fact that 1/3 of the global work force is unemployed or underemployed.

These three alternative approaches represent different trade-offs between the three pillars of sustainable development. To some extent, they represent unpackaging the system into component parts in order to make it manageable. However, if the problem is 'systemic and temporal' it cannot be addressed through atomistic and linear options (Gerlach, 1992: 71–3). Gerlach argues that solving global problems is not about picking the easy options but about restructuring society and facing the difficult questions. However, systemic change is getting more and more difficult to achieve as vested interests try to maintain the status quo and are only willing to change the system at the margins.

2.5 Dealing with underlying driving forces

2.5.1 Introduction

Apart from systemic issues, there are a number of key underlying driving factors that lead to large-scale GHG emissions. These include population growth and the continuous drive to produce, consume and trade. A European Commission (EC, 2012) report states that the global population may peak at 9.2 billion around 2075, life expectancy may increase to about 106 from around 2030 onwards, there will be regional imbalances in population concentrations and increased urbanization. Economic growth and enhanced global connectivity will lead to more resource use and greater pollution.

2.5.2 Options and issues

Each of these drivers calls for different sets of policy options. To the extent that production processes are responsible, there is need for life cycle management of products from 'cradle to grave to cradle' to ensure that 'material loops' are closed locally or globally. In other words, society should reduce, reuse and recycle resources. Dematerialization and decarbonization of these processes can ensure that fewer materials are used in production processes and that the carbon content of these production processes is minimized. Industry needs to be encouraged to follow the principles of corporate social responsibility and the Organisation for Economic Co-operation and Development (OECD) standards for self-regulation. To the extent that consumption is the key challenge, there is a need to influence this through procurement policies at state level, sustainability labelling and enhancing consumer knowledge in order to promote conscious behaviour and choices. To the extent that the problem is investment, there is a need to promote green investment approaches through screening bank investments and export credits. To the extent that trade encourages greater and unsustainable exploitation of resources, there may be a need for measures to promote sustainable development. Population growth can only be dealt with by empowering women and promoting sexual equality, health care and social security systems.

2.6 Dealing with proximate driving forces: mitigation and sequestration options

2.6.1 Introduction

Greenhouse gases are emitted mostly by the energy supply, transport, buildings, industry, agriculture and forestry and waste disposal sectors. For each of the above-mentioned sectors a range of possible measures can be taken.

2.6.2 Options and issues

Reducing GHGs from the energy supply sector calls for fuel switching from fossil fuels to low- or no-carbon energy sources such as renewable energy (or nuclear energy), improving technologies in the fossil fuel sector to reduce the rate of emissions in production and distribution, and combined heat and power. The transport sector can reduce its GHG emissions through focusing on transport switching from cars to bicycles or public transport, spatial planning, using cleaner fuels (e.g. gas, clean diesel, possibly biofuels), increasing fuel and transport efficiency, and hybrid vehicles. The building sector could benefit through better use of architectural design to reduce energy needs (e.g. insulation), better electrical appliances for heating, lighting and cooling, and alternative refrigeration fluids. Industry can become more efficient through better process technologies, heat and power recovery, dematerialization, decarbonization and better management systems. Agriculture can use better crops, land management, water management, rice cultivation, livestock and manure management, energy systems and can investigate whether biofuel generation can be reconciled with food production. The forest sector can benefit by reducing deforestation and enhancing afforestation. The waste sector can promote reduction, reuse and recycling of products, methane recovery from landfills, efficient waste treatment, and energy recovery from incineration processes (IPCC-III, 2007). Carbon capture and storage (CCS) may arguably play a role here in the future. This implies removing carbon from large sources of CO_2 (e.g. fossil fuel plants) and storing it in empty oil or gas fields or deep saline formations underground. In theory there is considerable potential for such storage. However, removing carbon itself is an energy-intensive process. The cost of storing the carbon in geological formations depends on the costs of separating the exhaust gas or the pre-combustion separation. The cost of capturing the carbon is the most expensive part (IPCC, 2005).

The technical potential for reducing GHG emissions is huge – almost twice as much as is needed for 450 ppm CO_2 equivalent stabilization (IPCC-III, 2007; Metz, 2010: 61). It is expected that the costs of these changes (abatement costs) could cumulatively amount to about 2–3% of global GDP for such tough stabilization levels (Metz, 2010: 69). The global costs of mitigation could be as high as USD 200–210 billion in 2030 (UNFCCC secretariat, 2008). These costs may be lower if technologies become cheaper through high deployment, as in the case of solar photovoltaics. Furthermore, many of these mitigation options have major co-benefits or additional benefits – reducing fossil fuel use can reduce the emissions of sulphur dioxide and nitrous oxide which affect human health. Energy efficiency in houses reduces the energy bills of households. New technologies may lead to new jobs. However, societies are locked into specific long-life infrastructures and many vested interests oppose change as they have a lot to lose from shifts to a more decarbonized lifestyle. Furthermore, country-specific costs are different and affected by fears of free-riders, leakage and loss of competitiveness (see 1.3.1).

2.7 Dealing with atmospheric concentrations and warming

2.7.1 Introduction

When GHGs accumulate in the atmosphere they increase in concentration levels; increased concentration levels could enhance warming. One way to deal with these aspects in the climate process chain may be through geo-engineering (earth system engineering) – 'intentional large-scale manipulation of the environment' (Keith, 2000: 247). Discussions about geo-engineering began in the early 1990s; the earliest discussions focused on carbon capture and storage (see 2.6). Over the last 20 years there have been recurring discussions of geo-engineering as an option, but in recent years the lack of progress made in the area of climate mitigation has once again given additional impetus to the climate change geo-engineering options.

2.7.2 Options and issues

Geo-engineering includes deflection of the sun's long-wave rays through reflection by large-scale space-based mirrors, reflective stratospheric aerosols, cloud albedo enhancement and surface albedo increases. The mirrors/lenses (either a giant space screen or small lenses (Angel, 2006)) would need to reflect 1.58–1.80% of the sunlight to compensate for a doubling of CO_2 concentrations (Angel, 2006; Lenton and Vaughan, 2009). Stratospheric aerosols include emitting sulphur dioxide/sulphate aerosols into the stratosphere so that they reflect the sun's rays back. Annually, 1.5–5.0 Gt sulphur would need to be emitted to compensate for a doubling of CO_2 concentrations (Crutzen, 2006; Rasch *et al.*, 2008; Salter *et al.*, 2008; Teller *et al.*, 2002). Cloud seeding through spraying with sodium chloride from the ocean (mechanical: Latham, 1990; Latham *et al.*, 2008), or through iron injections into the ocean leading to a higher mass of algae which emit dimethyl sulphide (biological: Charlson *et al.*, 1987; Wingenter *et al.*, 2007), can also reflect long-wave radiation. Surface albedo can also be enhanced by painting roofs white (Hamwey, 2007), and putting a white screen on deserts (Gaskill, 2004), crops and grasslands (selected for reflectiveness: Hamwey 2007).

Geo-engineering also includes the extraction of atmospheric GHGs through carbon capture from ambient air using energy-intensive large scrubbing plants, and biomass energy with carbon storage which involves afforestation for the purpose of growing energy plants (Read, 2008). Other options include advanced weathering (removing carbon through chemical reactions with carbonate and silicate rocks, and enhancing carbon uptake through biological activity in the oceans). Afforestation, reforestation, carbon capture and carbon storage may, in some definitions, also fall under the realm of geo-engineering (see 2.6).

There are a number of pros and cons for geo-engineering. Geo-engineering can be seen as an important emergency back-up strategy if other means don't work out, a relatively cheap mitigation option to hold off the worst possible effects of climate change (Wigley, 2006), and it may even incentivize mitigation measures (Virgoe, 2008). However, solar radiation management options (such as aerosols and mirrors) are essentially short-term, non-structural solutions, which may be an advantage as projects can be stopped if there are negative side effects. Second, their deployment requires a very good understanding of how the climate system works and how one can manipulate the system (Benford, 1997; Schneider, 1996). Because knowledge continues to be imperfect, such deployment could lead to major problems. Third, it addresses the symptoms of the problem and not the causes. Solar radiation management could disturb atmospheric circulation, increasing the risk of droughts and floods at some places instead of reducing it. Finally, the side effects of the interventions are not clear and may even have differential impacts across regions (e.g. one country could increase its precipitation at the cost of another (Ricke *et al.*, 2010)). It is also not entirely evident that these options are cheaper than others. Options focused on short-wave radiation are fast but temporary, while those focused on the long-wave radiation are slow-working but may have a longer-term effect. In the meantime, ocean acidification would increase unhindered, damaging ecosystems. If one were to persevere with solar radiation management for whatever reason, the temperature would increase dramatically in a very short time frame. Nor is it entirely clear whether these interventions are legally possible. The UN Convention on the Prohibition of Military or any Other Hostile Use of Environmental Modification Techniques (UN, 1976) does not allow interventions that can be environmentally damaging for military reasons. However, the law may not necessarily rule out the use of geo-engineering. The Convention on the Prevention of Marine Pollution by

Dumping of Wastes and Other Matter (IMO, 1972) and the UN Convention on the Law of the Sea (UNCLOS, 1982) may regulate the legality of ocean fertilization (Freestone and Rayfuse, 2008).

2.8 Dealing with impacts

2.8.1 Introduction

The impacts of climate change include rising temperatures, rising sea levels, melting glaciers and snow-capped regions and extreme weather events. This leads to an increase in glacial lakes, land instability in permafrost areas, increased runoff and change in discharge periods, ocean acidification and related changes in ecosystems such as early springtime activities, poleward shifts of species, and fish migrations. These impacts are expected to influence freshwater systems, changes in food, fibre and forest productivity, affect coastal areas, ports and small island states, and coral reefs, putting millions of people in these areas at risk. These have an impact on industry, settlements and society, in general, both directly and indirectly, through health impacts such as increased malnutrition, the spread of disease and the increased likelihood of death from heatwaves, floods, storms, droughts, fires and cardio-respiratory disease as a result of higher ground-level ozone. The impacts will vary in different parts of the world, but it is expected that hundreds of millions of people will be exposed to increased water stress, higher prices because of lower productivity of food in specific areas, and coastal flooding; coral bleaching is already occurring and up to 30% loss of species is expected at a 1–2° warming (IPCC-II, 2007, Summary).

2.8.2 Options and issues

It would perhaps be appropriate to briefly explain key relevant terms: vulnerability, adaptive capacity and resilience (the ability to bounce back after an impact). Adaptation can be a process or a structure. Anticipatory adaptation, as opposed to reactive adaptation, takes place before an impact. Autonomous, as opposed to planned, adaptation is triggered by an impact and planned adaptation is the result of a conscious decision; differentiating between the two may be difficult (Fankhauser *et al.*, 1999). Autonomous adaptation depends on many aspects including institutional support, manpower and financial and technological resources (see Ausubel, 1991; Mendelsohn and Neumann, 1999; Mendelsohn and Nordhaus, 1999; Yohe *et al.*, 1996). Adaptation can occur from local through to global levels, but tends to be more local in nature. It can occur in the short term through to the long term, and can involve the public and/or private sectors.

Vulnerability is the degree to which a system is susceptible to, and unable to cope with, adverse impacts of climate change, including climate variability and extremes. Vulnerability is a function of the character, magnitude, and rate of climate change and variation to which a system is exposed, its sensitivity and its adaptive capacity [italics removed].

(IPCC-II, 2007, Summary: 93)

Exposure refers to whether a region is likely to experience a disaster/ hurricane/sea level rise/drought. Sensitivity refers to the characteristic (level above sea, population density, existing temperature) of a region to a specific harm. These characteristics include the existing stresses to which a specific region may already be subject. Vulnerability can be exacerbated by the choice of a specific development path (e.g. tourism and construction in coastal regions can enhance coastal vulnerability).

Resilience is defined as 'the ability of a social or ecological system to absorb disturbances while retaining the same basic structure and ways of functioning, the capacity for self-organization, and the capacity to adapt to stress and change' (IPCC-II, 2007: Summary, 90). Adaptive capacity is often used interchangeably with resilience, but adaptive capacity describes the ability of the community and the ecosystem to prepare for, or cope with, such a natural or man-made disaster (cf. MA, 2005: 599). It is also closely linked to six factors: variety (of problem framing and solutions, of actors and sectors, and redundancy in the system); learning capacity (trust, single- to double-loop learning, the ability to discuss doubts, institutional memory); room for autonomous change (access to information, ability to improvise); leadership (visionary, collaborative, entrepreneurial); resources (financial, human, authority); and fair governance (legitimacy, accountability, equity, responsiveness) (Gupta *et al.*, 2010; Smit and Pilifosova, 2001; Smit *et al.*, 2001).

All societies need to adapt to the impacts of climate change. In the case of developing countries, they may need assistance to adapt. Here, critical questions are whether the adaptation is with respect to climate change or climate variability and to what extent the impacts are aggravated by local sensitivity. 'Nevertheless, many in the climate community are calling for a 'no-regrets' adaptation strategy that integrates adaptation to climate change with adaptation to 'normal' climate variability' (Linnerooth-Bayer and Mechler, 2006: 622). Adaptation aims to prevent risk, reduce vulnerability, and cope with specific climatic events.

Climate change and variability adaptation is already taking place in rich countries like the Netherlands (van der Grijp *et al.*, 2012) as well as in poor countries like the Maldives (IPCC-II, 2007: Summary). The major sectors affected by adaptation are water, agriculture, infrastructure and settlements, human health, tourism, transport and energy. Measures for these sectors include rainwater harvesting, storage and desalination techniques, water reuse and efficiency in irrigation and industry; micro-level options such as crop diversification, changing crop intensity,

mix and location; market responses such as crop insurance, pricing and trade policies; technological options such as forecasting; relocation, sea and river defences
and climate proofing infrastructure; improving water access and sanitation, disease
monitoring and prevention; emergency and normal health plans; diversification of
coastal and ski-tourism plans for water storage; climate-proofing transport and
energy systems (IPCC-III, 2007). Policy options include regulatory instruments
(targets, standards, monitoring, temporary or permanent organized migration, food
and water storage, land tenure, spatial planning to provide more room for nature
or deal with coastal squeeze as a result of sea-level rise), economic and financial
instruments (subsidies, taxes, to diversify income), and research and suasive instruments.

The costs of adaptation can be quite substantial for small vulnerable countries,
and quite small for large and more prosperous countries. Total global costs needed
for adaptation in 2030 could range from USD 49 billion up to 171 billion (at 2005
USD prices) (UNFCCC Secretariat, 2008). Different organizations have different
statistics on the resources needed for adaptation purposes.

2.9 Dealing with residual impacts

2.9.1 Introduction

There are always sudden and slow-onset impacts that cannot be easily dealt with
and have to be experienced (Linnerooth-Bayer and Mechler, 2006), and the effects
may have to be cushioned later, somehow.

It is estimated that global weather-related economic costs could reach USD
1 trillion annually by 2040 (Dlugolecki, 2008). Already in Europe, the costs
have reached EUR 8 billion over a 20-year period (Dlugolecki, 2008). Floods
have increased 15-fold on average per year in the period 1993–2001 (European
Environment Agency, 2004). The hot summer of 2003 caused 71,000 premature
deaths (Robine *et al.*, 2007 cited in Dlugolecki, 2008); it also led to major social
and economic disruption, including on inland shipping and nuclear power plants
(Munich Re, 2004 cited in Dlugolecki, 2008). The costs of the Katrina disaster in
New Orleans in 2005 and 2006, for example, were estimated at between USD 100
and 200 billion, while a programme constructing levees as proposed in 1998 would
have cost USD 14 billion (Fischetti, 2005: A23). The recent Sandy event of 2012 is
likely to cost somewhere in the range USD 30–50 billion. Similar events worldwide
also have high costs of residual impacts. It is argued that the cost of weather disasters has gone up since the 1950s by 5% per year (Dlugolecki, 2008, citing Miller
et al., in press, and UNEP, 2006). This is partly because of anthropogenic climate
change, but also because of increasing population and urbanization in vulnerable

areas (for an overview see Bouwer, 2011). The level of absolute economic losses from weather-related events is high for the large, rich countries; however, the impacts for the smaller countries often represent a very large percentage of their national income. For instance, losses in the Caribbean after the 2004 hurricanes ranged from 1.9% of GNP in the Dominican Republic to 183% in the Cayman Islands and 212% in Grenada (Hoeppe and Gurenko, 2006).

2.9.2 *Options and issues*

The three options for coping with residual impacts include insurance, temporary or permanent migration and liability. This section discusses insurance and migration. Liability is discussed in Chapter 9.

Insurance companies can insure residual impacts and can reimburse claims when such costs occur. About USD 600 billion is needed to cover weather-related losses today. The global insurance industry has only about USD 150 billion to cover these risks, but there are other sources of funding from government emergency mechanisms and private savings. However, there is a shortfall (Dlugolecki, 2008) and possibly 90% of the total climate-related disaster costs are borne by households, companies and governments (Hoeppe and Gurenko, 2006). This does not mean that insurance schemes need to be made more comprehensive.

Disaster insurance in various forms can be beneficial as it increases government responsibility and ownership of possible problems, saves lives and livelihoods, allows for long-term planning, diminishes some of the negative effects of ad hoc aid, reinforces state responsibility, ensures dignity for victims, and builds on a rights-based approach. More importantly, by creating awareness and also providing financial incentives, risk transfer and insurance products may promote actions on risk reduction and adaptation (Warner *et al.*, 2009).

However, the risk premium (plus transaction costs) may be too expensive, may conflict with other plans if not integrated into a broader contingency plan, may not address all impacts, can provide relief but not new jobs or market access, is difficult to assess monetarily, and it is not always easy to engage appropriate stakeholders (Pierro and Desai, 2008). Different kinds of insurance exist: for flood and water risks, agricultural risks, risks to infrastructure, health risks, energy and economic loss risks. Also, different models of insurance and risk transfer may be applied to directly serve the citizens or indirectly through governments (see Warner *et al.*, 2007).

In relation to developing countries, there were initially discussions of an international insurance pool that would cover slow-onset events, like a rising sea level, to cover the risks faced by small island states (AOSIS Protocol, 1994). Subsequently, suggestions were made to establish insurance for sudden impacts that would be financed not by ad hoc, voluntary payments as currently occurs for disaster miti-gation, but by structured funding to a Climate Impact Relief Fund which could be disbursed by the Climate Convention (Müller, 2002). Another proposal was for a Climate Change Funding Mechanism, a global catastrophe insurance scheme, that would be funded by developed countries and would cover the costs of damage to public infrastructure in the developing countries (Bals *et al.*, 2006). A two-tiered approach has also been recommended in the literature – where the first tier focuses on establishing a local to national climate insurance programme in developing countries and a second tier which uses adaptation funding for post-event relief for uninsured events (Linnerooth-Bayer and Mechler, 2006). Within the Munich Climate Insurance Initiative (www.climate-insurance.org), proposals for a com-prehensive insurance mechanism, including incentives for risk reduction are being developed in the context of the UNFCCC. Local options include micro-insurance and climate-indexed products, weather derivatives and catastrophe bonds. Climate-indexed products pay out if the weather experienced is beyond a specific range based on historical experience. It is argued that disaster safety nets are absolutely essential to cushion the poorest from the impacts of extreme weather events – for example, Hurricane Mitch which led to reduced GNP by 6% four years after the event (Linnerooth-Bayer and Mechler, 2006). The problem is raising the necessary resources to cover such events and the willingness of governments to deal with disasters, whether or not a climate causality can be established.

In the context of developed countries, there will also be need for insurance. The insurance and reinsurance companies have been monitoring climate impacts and weather-related events for quite some time. They have developed a number of cri-teria for insurability and a number of insurance products. They have then comple-mented these products by innovative tools such as the Carbon Disclosure Project, which requires member companies to disclose whether climatic impacts could influence their operations. The Investors' Network on Climate Risk requires com-panies to climate-screen their investments (Herweijer *et al.*, 2009). The insurance companies have been reluctant to increase their premiums based on expectations, but without increased premiums, paying out is problematic. In 1992, Hurricane Andrew caused damage calculated at USD 22.3 billion (in 2005 USD) leading to the insolvency of nine insurers (Herweijer *et al.*, 2009; cf. Vellinga *et al.*, 2001). Insurance companies face unprecedented events, a low likelihood of high pre-miums (if society does not see climate change as serious, why will they insure against it?), possibly high exposure, and low claim-handling capacity and expe-rience. The availability and affordability of insurance is also at risk (Herweijer *et al.*, 2009). The insurance industry is thus likely to push for climate mitigation to keep the impacts within limits. It is also likely to promote risk-based pricing, to require insurers to 'climate proof' their properties (e.g. build shutters in regions exposed to storm surges) and to offer risk education and advisory services and disaster resilience practices (Herweijer *et al.*, 2009). Insurance also presents a 'moral hazard' where those insured feel a reduced impetus to take action (Vellinga *et al.*, 2001).

Another option is to migrate. Environmental refugees are those who may have to relocate because of environmental problems. Climate change may require large-scale relocation of people from low-lying coastal areas, but also for those affected by long-lasting droughts and floods. Technically, these people do not qualify for refugee status under the United Nations Convention in relation to the Status of Refugees 1951. There is now increasing emphasis on the need to understand the causes of migration, how policies can be developed to support temporary and permanent migration, and how the international community will respond to the rising flow of environmental migrants (see also Chapter 9).

2.10 Structure and outline of this book

Having explained some of the key issues in relation to climate change, this book tries to take a bird's-eye view of the historical process of governance. By taking this overview it aims to rise above the individual challenges that each period has faced to see an overall picture of the strengths, weaknesses, opportunities and threats that face the regime.

As this book unfolds, it will become clear that the global climate change negotiating process has gone through five phases to date (see Table 2.1). In the first phase, the rise of scientific knowledge and political consensus led to a United Nations General Assembly decision to begin the process of negotiation (see Chapter 3). In the second phase, the negotiation process led to a definition of the underlying paradigm of northern leadership and the outline of an umbrella agreement (see Chapter 4). During the third phase, the momentum continued despite the fear that the USA and Australia might not participate in such a Kyoto Protocol; countries made their leadership conditional on others taking action (see Chapter 5). The fourth phase marked the rise of a number of new, possibly competitive, agreements and the rise of leadership competition (see Chapter 6). The current phase is marked by global recession on the one hand and the rise of new emerging economies on the other. This reduces the impetus of the North to lead and creates the inevitability of the need for leadership from the South in a world of changing geo-politics (see Chapter 7).

Part 3 then examines the evolving country coalitions and social actors in the international regime (see Chapter 8). It pays special attention to the issue of litigation and social justice (see Chapter 9). Part 4 examines the regime evolution from the perspective of learning processes (see Chapter 10).

Table 2.1 *The phases of climate negotiations within and outside the regime*

Period	The paradigm underlying the UNFCCC regime	Activities outside the regime
1: Before 1990	Framing the problem	Scientific reports and declarations, political declarations, cities and states becoming engaged
2: 1991–1996	Leadership articulated	NGOs becoming engaged, industry and economists becoming more aware
3: 1997–2001	Conditional leadership	Climate sceptics more prominent UN agencies become engaged
4: 2002–2007	Leadership competition	Political declarations (G20, G7), new agreements (APP, IPHE, etc.)
5: Post-2008	Leadership during recession: Focus on developing countries The search for new ideas	Ideas: Low-carbon society; green economy; transforming the energy system UK Climate Act 2008: Changes in national governance patterns (three five-year periods needed simultaneously, commits future goverments)

Source: Building further on Gupta (2010). This material is reproduced with permission of John Wiley & Sons, Inc.

2.11 Inferences

This chapter has argued that the climate problem is closely linked to development. It has defined a climate change process chain and identified the policy measures most suitable to different parts of the chain. It is increasingly clear that mitigation, adaptation and the fall-back option of geo-engineering are closely linked. Clearly, mitigation options need to be prioritized. However, mitigation options may sometimes be counter-productive (Gupta, S. *et al.*, 2007; Revesz and Livermore, 2008). The promotion of biofuels as a substitute for fossil fuels may lead to deforestation and food price volatility (Clapp, 2009). The more one mitigates the less one needs to adapt. However, although some argue that adaptation can reduce the benefits of mitigation (Buob and Stephan, 2011; Ebert and Welsch, 2011; Ingham *et al.*, 2005), this can be contested. There are limits to how far countries and peoples can adapt. These limits are set not just by the tipping points and irreversible climatic changes, but also by how specific regional impacts may put an entire race of people at risk. The more vulnerable countries will need adaptation and adaptation funding within a global governance process (Ayers and Huq, 2009; Denton, 2010; Huq *et al.*, 2004; Mace, 2005; Smith *et al.*, 2011).

Part Two

The history of the negotiations

3

Setting the stage: defining the climate problem (until 1990)

3.1 Introduction

Unlike many other issues, climate change, a very specific, scientifically complex problem, entered the international agenda through an abstract theoretical awareness of the problem rather than an actual experience of its consequences. In the 1980s, it moved rapidly from the scientific arena to the political arena, picking up concern from non-state actors on the way. It then developed a twin track – a track in which the scientific process of assessing information was institutionalized in the Intergovernmental Panel on Climate Change (IPCC), and a negotiating track through the establishment of the Intergovernmental Negotiating Committee (INC) under the auspices of the UN General Assembly to initiate negotiations on a climate treaty. This chapter explains the chronology of events, focuses on how the problem was defined, the role of the key actors, the key outputs and major trends in this period ending in 1990.

3.2 The chronology of events

The first driving factor for global governance in the climate domain was the role of individual scientists from the 19th century onwards in incrementally identifying various aspects of the problem (see Table 3.1). This culminated in the scientific consensus that climate change is a serious threat to humanity at the first World Climate Conference organized by the World Meteorological Organization (WMO) in 1979. The Conference Declaration stated:

It appears plausible that an increased amount of carbon dioxide in the atmosphere can contribute to a gradual warming of the lower atmosphere, especially at high altitudes … effects on a regional and global scale may be detectable before the end of this century.

(WCC, 1979)

Table 3.1 *Key events and outputs in Phase 1 of climate governance history*

Year	Activity/Output	Key message
1979	World Climate Conf. Declaration	Nations must urgently prevent man-made changes to climate
1985	Villach Conference	Increasing GHG concentrations will warm the global climate
1985	AGGG established	First cooperative framework for undertaking science
1985	Vienna Convention	On the ozone layer
1986	UNGA right to develop	Developing countries have a right to develop
1987	Montreal Protocol	To limit emissions of ozone-depleting gases (also GHGs)
1987	Brundtland Report	Climate change is interlocked with other global issues; waiting for certainty may make action too late
1988	Toronto Conference	Climate change serious; Developed countries should reduce CO_2 emissions by 20% in 2005 relative to 1988 levels
1988	IPCC established	Cooperative science institutionalized
1988	UNGA 43/53	Climate change is a common concern; endorses IPCC
1989	Ottawa meeting	Discussed legal principles including a liability protocol
1989	Hague Declaration	Mobilized heads of state of 22 countries on this issue
1989	Declaration of Brasilia	Environmental challenges result from industrial patterns
1989	Commonwealth Langkawi Declaration	Responsibility should be equitably shared: 'to achieve sustainable development, economic growth is a compelling necessity'
1989	Tokyo Conference	Climate change global problem calling for environmental ethics, reducing uncertainties and better links between science and policy
1989	Noordwijk Declaration	Developed countries should stabilize CO_2 emissions by 2000 with respect to 1990 levels; and provide assistance to developing countries
1989	Malé Declaration	Recognized serious impacts especially on small islands
1989	UNGA 44/862	Recognizes the importance of not causing harm to others and the need for new and additional resources
1989	World Conference on Preparing for Climate Change	Cairo Compact: use peace dividend for environmental security
1990	Second World Climate Conference	It is possible for developed countries to stabilize CO_2 emissions from the energy sector and to reduce these by at least 20% by 2005. Developing countries should use modern technologies: 'leapfrogging'

Table 3.1 (cont.)

Year	Activity/Output	Key message
1990	ECE Conference, Bergen	The precautionary principle should guide action
1990	Report of IPCC	In a business-as-usual scenario, temperature will increase by 1°C by 2030; stabilization of concentrations would require 60–80% cuts in CO_2 emissions
1990	EC target	Stabilization of CO_2 emissions in 2000 with respect to 1990 levels
1990	G7	Reconfirmed need for action
1990	UNGA 45/212	Establishes Intergovernmental Negotiating Committee (INC) to negotiate climate convention.

Source: Elaborating further on Gupta (2010c). This material is modified and reproduced with permission of John Wiley & Sons, Inc.

It concluded that '[t]he long-term survival of mankind depends on achieving a harmony between society and nature' (WCC, 1979). In 1985, a scientific meeting was held in Villach which led to the establishment of the Advisory Group on Greenhouse Gases (AGGG) by WMO, the United Nations Environment Programme (UNEP) and the International Council of Scientific Unions. This body was requested to assess the science on the GHG problem. The world population was approaching 5 billion at this time and carbon emissions were about 6 billion tonnes per year.

Meanwhile, the international community adopted the Vienna Convention on the Ozone Layer in 1985 and the Montreal Protocol on Ozone Depleting Substances in 1987 to deal with the problem of depletion of the ozone layer. The Montreal Protocol aimed to phase out the use of chlorofluorocarbons (CFCs), which incidentally are also GHGs. In 1986, a parallel process at UN level led to the adoption of the Right to Development by the General Assembly (see 1.4).

At a public hearing of the World Commission on Environment and Development (WCED), Irving Mintzer argued that '[t]he ultimate potential impacts of a greenhouse warming could be catastrophic' (cited in WCED, 1987: 175). In 1987, the WCED under the leadership of Mrs Brundtland published its seminal report on 'Our Common Future', which identified the role of climate in the context of other global issues (WCED, 1987). It also asked the key question:

How much certainty should governments require before agreeing to take action? If they wait until significant climate change is demonstrated, it may be too late for any counter measures to be effective against the inertia by then stored in this massive global system.

(WCED, 1987: 176)

This was followed by meetings in Hamburg and Toronto. At the Toronto Conference, 300 experts from 46 countries concluded that the world was in the midst of an 'unintended, uncontrolled, globally pervasive experiment' and called on the industrialized countries to reduce their CO_2 emissions by 20% by 2005 compared with 1988 levels and to phase out the use of CFCs by 2000 (Toronto Declaration, 1988). They recalled the Trail Smelter case between the USA and Canada (Trail Smelter Case, 1941) and Principle 21 of the Stockholm Declaration on the Human Environment (Stockholm Declaration, 1972), which emphasize that states must not cause harm to other states.

In 1988, the record drought that hit the Midwest and many other parts of the US led to crop damages of about USD 40 billion (UCS, 2009) and focused the USA's attention on climate change. That year, a scientific assessment programme – the IPCC – was established by WMO and UNEP. This body had three working groups – one to work on the science of, one on the impacts of, and one on the response strategies to climate change. In December 1988, in response to the request of the Government of Malta, the UN General Assembly adopted a resolution stating that climate change is a 'common concern of mankind' and 'should be confronted within a global framework' (UNGA, 1988, 43/53).

In January 1989, 80 international legal and policy experts met in Ottawa, Canada. They recommended an umbrella (framework) convention which would recognize the atmosphere as 'a common resource of vital interest to mankind' (#A3). This led to a corresponding state 'obligation to protect and preserve the atmosphere' (#A4) which implies that:

The sovereign right of States to permit in their territories or under their jurisdiction or control all human activities that they consider appropriate must be compatible (must conform) with their obligations to protect and preserve the atmosphere (#A5).

Among other articles, the experts recommended the development and transfer of technology to developing countries (#A15), prior notice of environmental impact assessment of planned activities (#A16) and a protocol on liability, compensation or other relief (#A19). They recommended a World Climate Trust Fund for promoting relevant activities in developing countries, financed by voluntary or assessed contributions from states, 'user fees' for activities causing climate change, and fines for violating the articles of the proposed convention (#B9). They stated that the industrialized countries have the primary responsibility to reduce their emissions and that:

Having caused the major share of the problem and possessing the resources to do something about it, the industrial countries have a special responsibility to assist the developing countries in finding and financing appropriate responses (#B4.10).

At the same time, they argued that the contribution of developing countries is likely to rise from 20% to about 50% and that it was important to engage them 'in a way that enhances, rather than diminishes, development prospects' (#B4.9). They concluded that climate change requires a global response which needs to be differentiated between developed and developing countries (#B5.1). The document records a range of policy options including those on addressing deforestation and population growth.

In March 1989, the first high-level meeting of 24 heads of state (the USA and the UK did not attend) was held at The Hague under the leadership of Dutch Prime Minister, Ruud Lubbers, and the French and Norwegian governments. The French had apparently initiated the process as a way to celebrate their bicentennial and wished to promote an environmental revolution, but in the end the Netherlands hosted the event. The Hague Declaration (1989) stated that:

Because of the nature of dangers involved, remedies to be sought involve not only the fundamental duty to preserve the ecosystem, but also the right to live in dignity in a viable global environment and the consequent duty of the community of nations vis-à-vis present and future generations to do all that can be done to preserve the quality of the atmosphere.

The Hague Declaration also focused on the 'special obligations' of the developed countries to assist developing countries. It argued that:

(d) The principle that countries to which decisions taken to protect the atmosphere shall prove to be an abnormal or special burden, in view, inter alia, of their level of development and actual responsibility for the deterioration of the atmosphere, shall receive fair and equitable assistance to compensate them for bearing such burden.

The Latin American and Caribbean countries adopted the Declaration of Brasilia (1989), which emphasized the perspective of indebted Latin American countries – focusing on who was responsible for the problem, the need to address the debt problem and the need to find a balance between economic development and environmental protection. A few months later, the francophone countries met and endorsed the Hague Declaration (1989). In June, the European Community Council decided to promote action on climate change, and in July the G7 declared that GHG emissions needed to be limited. The Economic Declaration of the seven heads of state or government of the European Communities also took note of the increasing environmental degradation and the need for action. In September, the Tokyo Conference on the Global Environment recognized climate change as a global problem, calling for better science–policy interface and new environmental ethics. It focused on policy responses in relation to, among others, energy efficiency and deforestation.

The Non-aligned Movement countries met in September, the Commonwealth nations in October and the Small Islands in November. The Langkawi Declaration (1989) of the Commonwealth countries recognized climate change as serious and submitted that 'The responsibility for ensuring a better environment should be equitably shared and the ability of developing countries to respond be taken into account' (#5). It also argued that:

To achieve sustainable development, economic growth is a compelling necessity. Sustainable Development implies the incorporation of environmental concerns into economic planning and policies. Environmental concerns should not be used to introduce a new form of conditionality in aid and development financing, nor as a pretext for creating unjustified barriers to trade (#6).

While the major meetings of the developing countries did not discuss climate change in particular, they discussed the disruption of the global ecological balance (NAM, 1989). The Malé Declaration (1989) of the Small Islands on Global Warming and Sea Level Rise urged all countries to reduce their GHG emissions. To cement the progress made, a meeting was convened for environmental ministers from 67 countries at Noordwijk in the Netherlands. Five background reports were prepared for this conference: one that linked climate change to other environmental and developmental issues, one on greenhouse gases, one on the elements of an international convention on climate change, one on forestry and afforestation, and one on funding mechanisms. This last document provided the early ideas for the instrument that later developed into Joint Implementation and the Clean Development Mechanism (see Chapters 4–7). Discussions on these documents and related deliberations led to the adoption of the Noordwijk Declaration on Climate Change. This declaration (Noordwijk Declaration, 1989: Art. 8) also provided initial language for the long-term objective of any future treaty on climate change. It recommended a stabilization target for CO_2 for 2000 and a reduction of CO_2 emissions by 20% by 2005 with respect to 1990 levels, net afforestation of 12 million hectares a year as of 2000 onwards, phasing out of CFCs and limiting the emissions of other GHGs (Noordwijk Declaration, 1989: Arts 14–23). The Noordwijk Declaration recognized the co-benefits of taking climate action – i.e. that many actions have other benefits over and above reducing GHGs (Noordwijk Declaration, 1989: Art. 5); that industrialized countries should set an example by initiating action to reduce their own emissions of GHGs; that they should support developing countries financially and otherwise to reduce their emissions, and that these actions should be undertaken recognizing the 'need of the developing countries to have sustainable development' (Noordwijk Declaration, 1989: Art. 7).

The UN General Assembly (1989a) reaffirmed that:

in accordance with the Charter of the United Nations and the principles of international law, States have the sovereign right to exploit their own resources in accordance with their environmental policies, and also reaffirms their responsibility to ensure that activities within their jurisdiction or control do not cause damage to the environment of other States or of areas beyond the limits of national jurisdiction and to play their due role in preserving and protecting the global and regional environment in accordance with their capacities and specific responsibilities (para. 4).

It also stated that:

owing to its universal character, the United Nations system, through the General Assembly, is the appropriate forum for concerted political action on global environmental problems (para 5).

And it:

[e]ncourages Governments and relevant international organizations to further the development of international funding mechanisms, taking account of proposals for a climate fund and other innovative ideas, bearing in mind the need to provide new and additional financial resources to support developing countries in identifying, analysing, monitoring, preventing and managing environmental problems, primarily at their source, in accordance with national development goals, objectives and plans, so as to ensure that development priorities are not adversely affected (para 14).

In December 1989, the World Conference on Preparing for Climate Change called for cooperation 'on an unprecedented scale', using the peace dividend 'to be deployed in the pursuit of environmental security instead'; called on all social actors to engage in addressing the problem; and listed policy measures including World Cultural Zones to safeguard national and global heritage (Cairo Compact, 1989).

1990 was a critical year. The Economic Commission of Europe met in Bergen and adopted the 'precautionary principle'. At Bergen, as at Noordwijk, the opposition of the USA, Japan and the Soviet Union to quantitative targets led to a weaker wording of the target to bring back emissions of CO_2 by industrialized countries by 2000 at 1990 levels (Bodansky, 1993). The G7 Houston meeting reconfirmed the commitment of the world's most powerful countries to take action. The European Community (1990), consisting of 12 countries at the time, agreed to 'take actions aiming at reaching stabilization of the total CO_2 emissions by 2000 at 1990 level in the Community as a whole'. The Second World Climate Conference adopted a scientific statement and a political statement (SWCC, 1990). The political statement recognized the concept of climate change as a common concern of humankind and adopted the principle of equity, and the common but differentiated responsibilities of countries at different levels of development in dealing with climate change. It adopted the principle of sustainable development, and the idea that states should adopt measures even if the cause–effect science was uncertain, if the possible impacts of a problem were likely to be irreversible – the precautionary principle. It urged developed countries to take urgent action also in order to make space for the emissions of developing countries that will need to grow to meet their development needs.

The IPCC came up with its first assessment reports on science, impacts and policy. It predicted that under a business-as-usual scenario, global mean temperatures were likely to rise by about 0.3°C per decade amounting to a 3°C increase ('this is greater than that seen over the past 10,000 years' (IPCC-I, 1990: xi) and a sea level rise of about 65 cm by the end of the century. It explained some of the key uncertainties and the need for action. IPCC-II (1990) discussed the possible impacts on different sectors of society. IPCC-III (1990) elaborated on the possible responses in four sectors – energy and industry; agriculture, forestry and other human activities; coastal zone management; and resource use and management. It suggested the need for significant public education and information, technology development and transfer, economic measures, financial mechanisms and legal and institutional mechanisms. It recommended that a framework convention should discuss climate change as a common concern, and that countries should be able to exploit their natural resources but have a concomitant duty to protect and preserve the climate. It stated that there was need for:

[R]ecognition of the responsibility of all countries to make efforts at the national, regional and global levels to limit or reduce greenhouse gas emissions and prevent activities that could adversely affect climate, while bearing in mind that:

- Most emissions affecting the atmosphere originate in industrialized countries where the scope for change is the greatest;
- Implementation may take place in different time frames for different categories of countries and may be qualified by the means at the disposal of individual countries and their scientific and technical capabilities;
- Emissions from developing countries are growing and may need to grow in order to meet their development requirements and thus, over time, are likely to represent an increasingly significant percentage of global emissions.

(IPCC-III, 1990: 263)

The year ended with the UN General Assembly launching the Intergovernmental Negotiating Committee (INC) to negotiate a framework convention on climate change (UNGA, 1990). This Committee was set up under the auspices of the UNGA but was to be supported by UNEP and WMO. Developing countries supported the choice for UNGA as climate change was seen as a political and not a technical issue (Bodansky, 1993). The framework convention was to include appropriate commitments that took the work of the IPCC and the Second World Climate Conference into account and was to be ready by the 1992 Rio Conference on Environment and Development. The convention was to be open to all states; all UN organizations, and members of the scientific community, industry, trade unions and non-governmental organizations were also invited. The negotiations were to be funded through existing UN resources and through voluntary contributions to a fund. Funds were also to be made available to help the least developed and small island states to participate in the process.

Within this decade the scientific community was mobilized and organized, the UNGA took action and world leaders started drawing up a blueprint for action (see Table 3.1). This chapter takes you through the events of the early days.

3.3 The problem definition and measures discussed

3.3.1 The problem definition

The climate problem was defined as a very serious problem affecting human survival (WCC, 1979), second only to a nuclear war (Toronto Declaration, 1988), and was ultimately phrased as a global 'common concern' of humanity (Noordwijk Declaration, 1989; UNGA, 1988). There were few sceptics at this point of time (cf. Bodansky, 1993).

The problem of climate change was not questioned but framed in abstract terms as a global, future problem, which gave it a remote character. The local and immediate nature of the problem was not emphasized. The framing was essentially technocratic in nature – focusing on the sources of emissions and removals by sinks. There was some debate on how to deal with the uncertainty – while the Tokyo Declaration called for more clarification on the issue of uncertainty, the Brundtland Commission questioned whether more certainty was needed before action was taken (WCED 1987).

The problem called for stabilizing concentration levels (Noordwijk Declaration, 1989). IPCC submitted that in the event that states decided to stabilize GHG concentrations at the 1990 level, this would require that global GHG emissions would have to be reduced substantially: CO_2 by 60%, methane by 15–20%, nitrous oxide and CFCs by about 70–85% (Houghton *et al.*, 1990: xviii). This set, to some extent, the tone for the discussion with respect to the long-term objective of the future climate change convention.

Already by 1989 there was growing awareness that this problem had a 'North-South' character. It was argued that:

[r]elative to populations, cumulative [from 1950] historic fossil carbon emissions from industrialized countries are eleven times as high as those from developing countries. If the remaining global fossil carbon budget were shared according to strict person-year equity, including historic emissions, industrialized countries have no emissions left.

(Krause et al., 1989: Executive Summary)

In order to solve this problem, the authors suggested allocating a carbon budget of 150 billion tonnes between countries. This would necessitate a carbon emission reduction by the developed countries of 20% by 2005, 50% by 2015 and 75% by 2030, compared with 1985 levels. In return, developing countries would have to ensure that their emission growth rate did not exceed 50–100% of 1985 levels (Krause *et al.*, 1989: 1.6–26). Scholars argued that a central issue for developing countries would be 'how to obtain compensation for the opportunity costs they will incur from foregoing or altering their development' (Zaelke and Cameron, 1990: 283). At the time, there was little differentiation between East and West.

3.3.2 Measures discussed

At this early stage in the discussions, elements of a solution were already emerging. The Toronto Declaration (1988) had set a strong interim target. The Noordwijk Conference report suggested a long-term objective, interim targets for emission reductions and net afforestation, funding mechanisms and the development and transfer of appropriate technologies, elements of a climate change convention, and the need to share responsibilities between countries based on their contributions to the problem and capabilities. The Second World Climate Conference Scientific Declaration (SWCC, 1990) further suggested that developing countries should bypass the development path of the developed countries by investing in modern technologies.

IPCC-III (1990) outlined a number of key ingredients needed to design a solution. It argued in favour of a global multilateral solution, where all countries have a common responsibility towards dealing with the problem. However, the industrialized countries had special responsibilities in terms of reducing their own emissions, helping developing countries, and making room for the emissions of developing countries to grow. The report recommended that states should promote sustainable development emphasizing that: 'It is imperative that the right balance between economic and environmental objectives be struck' (IPCC-III (1990: xxvi). It argued in favour of emission limitation or mitigation strategies and adaptation strategies. The former includes energy efficiency, clean energy, better forest management, phase-out of CFCs, better livestock management and sustainable agricultural management. The latter includes emergency and disaster preparedness policies, comprehensive management plans and improved efficiency of natural resource use. There were expectations that climate change could severely affect people. An early article argued that there may be up to 10 million environmental refugees (Jacobson, 1988: 39). Tolba (1989: 25) expected that about 50 million people could become environmental refugees. The IPCC report also recommended that there was a need to better inform the public about climate change, promote science, new technologies, behavioural and structural change and expand monitoring systems.

The first assessment report of IPCC-III (IPCC-III, 1990: lvii–lxii) suggested items that could be included in a treaty. This included elements for a preamble, definitions, general obligations, the organizational set-up of the treaty, items in relation to research, observation, analysis and exchange of information, development and transfer of technologies, the settlement of disputes and other provisions. In doing so, it recognized the need to avoid causing harm to other countries, to see climate change as a common concern, to endorse sustainable development, and to acknowledge the different responsibilities of countries. It also discussed whether a reference to the precautionary principle should be made, whether references to weather modification agreements should be made, and whether specific actions should be taken to deal with vulnerable countries. Curiously, the report also raised two questions: 'Should mankind's interest in a viable environment be characterized as a fundamental right?' and 'Is there an entitlement not to be subjected, directly or indirectly, to the adverse effects of climate change?' (IPCC-III, 1990: lviii, see Chapter 9). The IPCC report also examined the possible role of tradable emission permits, emission charges, subsidies and sanctions.

3.4 The role of actors

3.4.1 Individual countries active: an agenda is born

In the early days, some developed countries felt the need to table the issue of climate change and secure the commitment of other countries. Austria, Germany, Canada and The Netherlands were amongst the first to show some initiative in this respect. In 1988, the Netherlands adopted a national target to stabilize CO_2 emissions in 2000 with respect to 1990 levels (VROM, 1989). One year later, it had strengthened its commitment to stabilization by 1993–1995 and a 3–5% reduction by 2000 (VROM, 1990). It tried to collate existing and new policy measures as a first step towards outlining the elements of a national climate change policy. In 1990, the European Community of 12 countries had a joint stabilization target and soon thereafter the European Free Trade Agreement countries adopted a stabilization target (see Table 3.2).

The first documented comparative position of other countries on climate change can be seen in the report of the Noordwijk Conference on Climate Change. Policymakers from the 68 participating countries were forced to think about climate change and about their position. The USA discussed the substantial budget that went into the development of climate change science, the seriousness of the problem and the need for sound science. It recognized the 'special problems' of developing countries and stated: 'We fully endorse, and even share, the aspirations of the developing world for economic development' (Vellinga *et al.*, 1990: 117). The Union of Socialist Soviet Republics stated that, in the absence of reliable scientific knowledge, countries should be cautious about large-scale control measures (Vellinga *et al.*, 1990: 42–3). China recognized the problem and called for an

Table 3.2 *Country targets in 1990*

Group	Target	Country	Target
EC	Stabilize CO_2 emissions in 2000/1990	Belgium	Reduce CO_2 by 5% in 2000/1990
		Denmark	Reduce CO_2 by 20% in 2005/1988
		France	Stabilize per capita CO_2 at 2 tonnes by 2000
		Germany	Reduce CO_2 by 25% in 2000/1987
		Greece	Supports EC target
		Ireland	Supports EC target
		Italy	Stabilize CO_2 in 2000/1990
			Reduce CO_2 by 5% in 2000/1990
		Luxembourg	Supports EC target
		Netherlands	Stabilize CO_2 in 1994–5/1989–90
			Reduce CO_2 by 3–5% in 2000/1990
		Portugal	Supports EC target
		Spain	Supports EC target
		UK	Stabilize CO_2 in 2005/1990
European Free Trade Association	Stabilize CO_2 emissions in 2000/1990	Austria	Reduce CO_2 by 20% in 2000/1988
		Finland	Stabilize CO_2 in 2000/1990
		Iceland	Stabilize GHGs in 2000/1990
		Norway	Stabilize CO_2 in 2000/1989
		Sweden	Supports EFTA target
		Switzerland	Stabilize CO_2 in 2000/1990
Others	No common target	Australia	Stabilize non-CFC GHGs in 2000/1988
			Reduce non-CFC GHGs by 20% in 2005/1988
		Canada	Stabilize non-CFC GHGs in 2000/1990
		Japan	Stabilize CO_2 per capita in 2000/1990
			Effort to stabilize CO_2, methane and other GHGs in 2000/1990
		New Zealand	Reduce CO_2 by 20% in 2000/1990
		US	Annual afforestation of 1 billion hectares

Source: Based on Wolters *et al.* (1991)

equitable approach to addressing it (Vellinga *et al.*, 1990: 46–7). Brazil emphasized North–South asymmetry, India the need for developing countries to grow, Tanzania for a 'North–South bargain', Saudi Arabia called for new values linked to the concepts of global community and the new world order, Nigeria called for the need to revaluate current models of development, and the Philippines for the concept of burden sharing (Vellinga *et al.*, 1990: 56–9; 73–4; 67–8; 35–7; 81–3; 103–6, respectively).

3.4.2 *Science is institutionalized: an epistemic community is born*

Individual climate scientists working on a range of issues coordinated initially in the context of the WMO. Engineers and policy scientists became engaged in the Advisory Group on Greenhouse Gases and a broader range of scholars became engaged in the first IPCC report. The goal in those days was partly assessment and partly constructive engagement in generating ideas about how the challenge of climate change could be addressed. While the AGGG included primarily natural scientists and engineers, the IPCC gradually began to include a larger spectrum of researchers representing different disciplines. These early IPCC documents argued that there was a serious climate problem and that if GHG atmospheric concentrations were to be stabilized, this would require emission reductions of 60–80%, that the impacts would be substantial and that there was need for action. They even laid out elements of a draft treaty.

3.4.3 *Other social actors*

The initial political process mobilized non-governmental actors. They not only began to focus on climate change, but following past experiences, rapidly organized themselves into a coalition – the Climate Action Network – in 1989. At the Noordwijk Conference, the NGOs from 22 countries argued that industrialized countries should reduce their CO_2 emissions by 20% in 2000 with respect to 1988. In these early days, cities were also engaged in discussing climate change; Toronto and Melbourne were two such cities. Initial activism from cities led to the establishment of the International Council for Local Environmental Initiatives in 1990. In this first phase, the public was scarcely, if at all, engaged in the developed and developing world. There was little media coverage except around James Hansen's claim in the US Senate of global warming in 1988 and Thatcher's (1988) speech to the Royal Society.

3.5 The governance outputs

3.5.1 *The discourses: liability and the leadership paradigm*

A key issue in climate change governance was, and is, allocating responsibilities to countries for taking action. Initial discussions focused on who was to be held liable for climate harm. One of the earliest scientific articles on climate law explained that a key challenge:

is the difficulty of defining and determining concepts of liability, responsibility, and illegality for ensuring adequate compensation for the measurable harmful impacts of global warming.

(Zaelke and Cameron, 1990: 251)

The authors recommended two legal responses – the use of judicial processes building upon the 'no-harm' principle already being developed in international legal cases (Corfu Channel Case, Lac Lanoux Arbitration and the Trail Smelter arbitration) and in international declarations (The Stockholm Declaration, 1972; UN Charter, 1945), and/or developing a treaty on climate change. A Toronto meeting of legal and policy experts also mentioned liability and compensation as one of the issues that would need to be taken into account (Toronto Statement, 1989). The liability paradigm would imply that those countries that caused substantial harm to other countries would be held responsible for reducing their emissions; this would also implicitly provide notice to other countries that if their emissions went up, this could lead to harm and that they would in the future also be held liable for climate harm. This framing would lead to emission reduction in the large emitters and would ensure that countries started to decarbonize. It would also raise, through compensation, resources for those countries affected by climate change.

Gradually, the early discussions gave rise to the notion of the leadership paradigm. This framed developed countries as leaders rather than polluters. Leaders would show the way for carbon reform. They would not be obliged to compensate other countries, but instead could make resources voluntarily available to countries that were affected by climate change. Leaders would lead, the rest of the world would follow. There was no need for finger pointing or unpleasant liability discussions. This idea was slowly articulated in the Noordwijk Declaration (1989), the ECE Conference Statement (Bergen Declaration, 1990, para. 11), the Tata Conference Statement (1989: para. 4.10) and the Second World Climate Conference (SWCC, 1990: para. 5). The concept of leadership replaced the concept of liability and there was increasing consensus on this issue. This would have three elements – emission reduction by the industrialized countries, financial and technological support for the developing countries with excessive burdens, and the need to make space for the developing countries to develop (Noordwijk Declaration, 1989: para. 7; see also IPCC-III, 1990: xxvi).

3.5.2 The principles

The Brundtland report (WCED, 1987) argued for the further development of a number of legal principles. These included a principle on a fundamental human right to an adequate environment, intergenerational equity, prior environmental assessments, prior notification, access and due process, reasonable and equitable use of resources, prevention and abatement, strict liability, non-discrimination and equal access and treatment to all who participate in decision making. It stated that:

21. States shall cease activities which breach an international obligation regarding the environment and provide compensation for the harm caused.

The Tokyo Conference on the global environment and human response toward sustainable development (1989) talked of the need for new Environmental Ethics but did not develop this further. The SWCC Conference (SWCC Ministerial Declaration, 1990: para. 5) linked the leadership concept to the principle of the Common but Differentiated Responsibilities (CBDR) concept:

Recognizing further that the principle of equity and the common but differentiated responsibility of countries should be the basis of any global response to climate change, developed countries must take the lead. They must all commit themselves to reduce their major contribution to the global net emissions and enter into and strengthen co-operation with developing countries to enable them to adequately address climate change without hindering their national development goals and objectives.

The Bergen UNECE Conference (1990) articulated the precautionary principle in relation to climate change – stating that:

where there are threats of serious or irreversible damage, lack of full scientific certainty should not be used as a reason for postponing measures to prevent environmental degradation.

(Bergen Declaration, 1990)

This was somewhat reworded at the Second World Climate Conference as:

where there are threats of serious or irreversible damage, lack of full scientific certainty should not be used as a reason for postponing *cost-effective* measures to prevent environmental degradation [my italics].

(SWCC, 1990: 7)

The Alliance of Small Island States (AOSIS) argued that: 'For us, the precautionary principle is much more than a semantic or theoretical exercise. It is an ecological and moral imperative. We do not have the luxury of waiting for conclusive proof ... The proof, we fear, will kill us' (cited in Wolters *et al.*, 1991).

3.5.3 The long-term objective

A long-term objective was articulated in the Noordwijk Declaration (Art. 8) which is very similar to that eventually adopted in the UNFCCC. It stated that:

For the long-term safeguarding of our planet and maintaining its ecological balance, joint effort and action should aim at limiting or reducing emissions and increasing sinks for greenhouse gases to a level consistent with the natural capacity of the planet. Such a level should be reached within a time frame sufficient to allow ecosystems to adapt naturally to climate change, to ensure that food production is not threatened and permit economic activity to develop in a sustainable and environmentally sound manner. Stabilizing the atmospheric concentrations for greenhouse gases is an imperative goal.

3.5.4 Targets and timetables

In order to divide the carbon budget (see 1.4) between countries, three issues were critical at this time: there was a need to identify the nature of the targets and time-tables with respect to the reduction of GHGs, the specific level(s) of the targets and the issue of target sharing between countries.

The discussion on reduction targets and timetables in the pre-1990 period focused on a gas-by-gas approach versus a comprehensive approach, which included the major GHG gases. At that time, the European Communities were in favour of a gas-by-gas approach as a distinct part of a comprehensive approach. The advantages of a gas-by-gas approach were that countries could focus on individual gases – especially CO_2 – and try to work on reductions there. Industrial emissions of CO_2 were easier to measure. The advantage of a comprehensive approach was that it would allow individual countries to choose their own targets based on an assessment of their own national interests. This could allow them to make cost-effective trade-offs between emission reductions of the individual gases.

In terms of specific targets, the Toronto Declaration and the Noordwijk Declaration called for targets to stabilize the emissions of CO_2 and reduce emissions by 20% in the year 2005 with respect to 1988. Furthermore, in the Noordwijk Declaration on Climate Change (Noordwijk Declaration, 1989) the participating ministers agreed:

to pursue a global balance between deforestation on the one hand and sound forest management on the other. A world net forest growth of 12 million hectares a year in the beginning of the next century should be considered a provisional aim.

This was followed by the adoption of a CO_2 stabilization target in 2000/1990 by the EC and a similar target was adopted by the European Free Trade Association (EFTA) countries. As a consequence, individual countries started to make policy commitments regarding their emissions. By the end of 1990 all the 26 Organisation for Economic Co-operation and Development (OECD) countries had targets, barring Turkey and the USA (see Table 3.2). Soon thereafter, Poland and Hungary also agreed to stabilize their emissions of CO_2.

3.5.5 Policies and measures: winners and losers and no-regrets measures

In the early days there was corridor talk of winners and losers – whether extremely cold countries could benefit from warming and would thus have no interest in dealing with the challenge of climate change. It soon became politically incorrect to talk about winners and losers as this would involve speculation, and because most countries would lose in the long term if the non-linear impacts of climate change were to become evident, even though it was possible to think of people and sectors winning in the short term.

Clearly, the uncertainty of the climate change problem was a major bottleneck in adopting expensive measures. This led to the birth of the 'no-regrets' concept, the need to adopt measures that were beneficial in their own right and led, in addition, to benefits for climate change mitigation.

In the meanwhile all the scientific and policy declarations and the IPCC-III (1990) report were focusing on the various areas in which policies and measures would have to be taken in order to reduce GHG emissions. The IPCC recommended limitation strategies such as improved energy efficiency, cleaner energy sources, improved forest management, phasing out of CFCs, improved livestock and agricultural management, and adaptation measures such as emergency and disaster preparedness policies, comprehensive management plans, control of desertification and enhancement of crop adaptability.

3.5.6 Technology transfer – 'leapfrogging' and finance – 'new and additional'

Almost all early political documents mention the need for new technologies to help reduce the growth of emissions. This was in line with the prevalent conceptual visualization that the developed countries could easily decouple their GHG emissions from their economic growth, and it made sense for them to market modern technologies to the developing world so that they could leap-frog ahead (SWCC, 1990). This was also subsequently emphasized in the IPCC report. It is important to note here that these early documents do not refer to the history of 'white elephant' technology transfers (technology transfers of inappropriate technologies to countries) that had been discussed extensively since the 1950s in the world of technology transfer and alternative development.

Another key issue was the sensitive matter of finance. It was clear that resources would be needed for mitigation and adaptation. The question was: How should this be financed? This discussion was running parallel to other discussions that were taking place in the UN relating to development cooperation. In the latter discussions, there was a focus on ensuring that developed countries kept to their voluntary promise of delivering 0.7% of their national income as official development assistance (ODA). Only a handful of the developed countries had met this target. In the climate discussions, the focus was both on using ODA for these purposes (e.g. Tokyo Declaration) as well as on the need for 'new and additional' resources, over and above the 0.7% amount, to meet the commitment on climate change and other environmental issues. Hence, the early documents (e.g. European Community, 1990) include the term 'new and additional'. At the Non-aligned Movement meeting, Rajiv Gandhi proposed a Planet Protection Fund which would amount to about 0.1% of GNP of all countries except the least developed. The IPCC-III (1990: 254) report explained that while some countries saw the need for 'new and additional' funding, others did not see this as being feasible at the time.

3.6 Key trends in Phase 1

This initial period was characterized by enhanced scientific knowledge, the recognition of a looming serious problem that challenged society, the mobilization of politicians to examine the issue, the gradual engagement of the small island states, the articulation of a number of key ideas to address the problem, and a focus on practical sectoral solutions for mitigation and adaptation. Policy and politics was being driven by science, where science appeared to have an independent influence on politics. Furthermore, the climate change problem was seen as so serious that even though the UN General Assembly (UNGA) endorsed that the science could be left to UNEP and WMO, the issue of negotiating a solution was upgraded to the General Assembly itself. In other words the highest UN agency was charged with setting up the negotiating committee to negotiate a treaty on climate change. The initial ideas on how the problem was to be addressed also had a very high degree of idealism and did not necessarily reflect the emerging trend of neo-liberal, neo-conservative approaches in other fields of water and energy.

4

Institutionalizing key issues: the Framework Convention on Climate Change (1991–1996)

4.1 Introduction

The fall of the Berlin Wall in 1989 and the start of an East–West reconciliation process led to renewed global optimism. People expected a 'peace dividend' – money that could be dedicated to environmental issues. The successful negotiations on the depletion of the ozone layer provided inspiration as these demonstrated that anthropogenic emissions *can* have global impacts and that international cooperation *was* possible to address such a problem. Furthermore, the necessity of linking environmental and developmental issues was on the agenda of the 1992 UN Conference on Environment and Development (UNCED) at Rio, 20 years after the 1972 UN Conference on the Human Environment in Stockholm. This optimism characterized the first half of the decade.

The first phase (see Chapter 3) had set the stage for the climate negotiations. The Intergovernmental Panel on Climate Change (IPCC)'s first scientific report was just being digested by policymakers. Political declarations had crystallized the key issues. The UN Intergovernmental Negotiating Committee (INC) was channelling energies towards negotiating a treaty on climate change and promoting its entry into force. The signs were propitious.

4.2 The chronology of events

In 1991 the INC meetings started the climate negotiations under the leadership of Jean Ripert of France. It met in February 1991 and set up one working group on commitments to reduce emissions and one on implementation mechanisms. It discussed whether targets should focus on individual GHGs (gas-by-gas) or on all GHGs (comprehensive), and on whether text on 'new and additional' funding to finance implementation by developing countries (see 3.5.6) should be included. INC2 compiled a treaty proposal based on existing proposals and IPCC ideas, but was eventually withdrawn and a new version circulated. The USA continued to oppose the idea of 'targets and timetables' for developed countries. The Japanese proposed a compromise of 'pledge and review' which would allow countries to make voluntary pledges subject to a transparent review process. In June 1991, the Alliance of Small Island States (AOSIS, 1991) under the leadership of Vanuatu proposed an international insurance pool to compensate small island states for future climatic impacts. At INC3, the European Community supported 'targets and time-tables' and that approach was then adopted. During INC4, developing countries proposed a set of principles and small island states called for far-reaching targets for the developed countries. INC5 did not make much progress and the negoti-ating text was heavily bracketed. A 'square-bracketed' text refers to elements on which there is no consensus. This led to an extended bureau meeting. Here, the two disagreements between developed and developing countries on principles and tar-gets for the developed countries were resolved through textual editing (construct-ive ambiguity)! The second half of INC5 resolved the remaining disagreement about locating the financial mechanism at the Global Environment Facility (GEF), by including the word 'interim'. This disagreement arose from a strong devel-oped country belief that the GEF established by the World Bank, United Nations Environment Programme (UNEP) and United Nations Development Programme (UNDP) could manage the financial issues arising under the convention, and devel-oping-country distrust of an environmental fund located under the auspices of the World Bank (Gupta 1995). There remained some stress regarding the exclusion of the developing countries' 'right to development' (Bodansky, 1993).

INC6 in May 1992 adopted the United Nations Framework Convention on Climate Change (UNFCCC) (see 4.3.2), which was opened for signature at UNCED in June. The rapidity with which the convention was drawn up and adopted is remarkable given the scientific complexity, the high economic stakes, the perceived abstract nature of the future impacts and the differing interests of the global com-munity.

Normally, when a convention is adopted, countries wait until that convention enters into force before taking further action. However, countries decided not to wait for the entry into force of the convention and to continue the INCs until such time that the convention entered into force.

Parallel to the climate negotiations, there were negotiations on a biodiversity convention (CBD, 1992); the two conventions avoided overlapping conversa-tions. UNCED was also expected to lead to a forest convention; and so forests were down-played within the INC negotiations. During the climate negotiations, a deal was made to remove desertification and land degradation from the purview of climate change and discuss these in a different convention. Furthermore, GHG emissions from the international aviation and maritime transport sectors were left to the International Civil Aviation Organization and the International Maritime Organization, respectively. This would ensure that key issues were managed by the appropriate UN agencies.

The UNCED conference also adopted the Rio Declaration on Environment and Development (Rio Declaration, 1992), which included 27 principles to guide global through to national action. These principles called on states to exercise sovereignty but not cause harm to other countries (#2), promote equity and sustainable development (#3), integrate environment into development (#4), eradicate poverty (#5), support vulnerable countries (#6), recognize the common but differentiated responsibilities and respective capabilities of states (#7), eliminate unsustainable production and consumption patterns (#8), promote capacity building in and technology transfer to developing countries (#9), enact environmental legislation (#11), develop liability and compensation rules (#13), adopt the precautionary approach of taking measures even when the science is uncertain, if the possible outcome is irreversible (#15), apply the 'polluter pays' principle (#17) and notify others of disasters (#18) and planned activities (#19). Furthermore, forest-related principles were also adopted (Rio Forest Principles, 1992). Another product of UNCED – The Convention on Biological Diversity (CBD, 1992) – aims to conserve biological diversity, promote the sustainable use of its components and the fair and equitable sharing of the benefits arising out of the utilization of genetic resources (Art. 1). UNCED also adopted 'Agenda 21', with 40 chapters among which one focuses on protecting the atmosphere. Although the UNCED meeting aimed at creating a global consensus and a North–South compact, US President Bush stated that he was unwilling to compromise on American lifestyles and this was seen as casting doubt on whether the developed countries were serious about actually reducing their environmental footprint to make space for the economic growth of the South (Akumu, 1994).

In 1994, the Convention to Combat Desertification (UNCCD, 1994) was adopted. This convention aimed to deal with desertification and mitigate the effects of drought in order to achieve sustainable development.

That same year, the Climate Convention entered into force leading to the start of the meetings of the Conference of the Parties (COPs). In 1996, the IPCC's Second Assessment Reports showed renewed evidence of a 'discernible human influence on the global climate' (IPCC-I, 1996: 4) justifying action. Three GHGs had 'grown significantly': by 30% for CO_2, 145% for CH_4 and 15% for N_2O between 1750 and 1992 (IPCC-I, 1996: 3). Global mean temperatures were likely to rise by about 1.0–3.5°C by the end of the 21st century, higher than in the previous 10,000 years, and the sea level could rise by about 0.5 m (IPCC-I, 1996: 6). The impacts of global warming would probably be higher in the developing countries as these were more vulnerable because of their lower adaptive capacity (IPCC-II, 1996: 5; see Table 4.1).

Table 4.1 *Key events and outputs in Phase 2 of climate governance history*

Year	Activity/Output	Key message
1991–1992	INC 1–6	Long-term stabilization objective, policies and measures, mechanisms
1991	AOSIS proposal	International Insurance Pool
1992	UNCED: Rio Declaration	27 principles
	Rio Forest Declaration	Principles for forest protection
	Agenda 21: Atmosphere	Protection of the Atmosphere
	Biodiversity Convention	Protection of Biodiversity
1992–1995	INCs	Prepare for the first COP
1993	Bill Clinton becomes USA president	Political goal to stabilize US GHGs at 2000 with respect to 1990 levels
1994	Desertification Convention	To minimize desertification
	UNFCCC enters into force	Countries are obliged to implement
	AOSIS submits draft protocol	Need for CO_2 reduction targets of 20% by 2005 with respect to 1990 levels
1995	COP 1, Berlin	21 decisions including: AGBM Process for targets and timetables to be adopted at COP-3, pilot phase on AIJ, no new commitments for DCs; 1 resolution
1996	IPCC (SAR)	'The balance of evidence suggests a discernible human influence on global climate'
	COP 2, Geneva	17 decisions including: Discussing targets for developed countries; 1 resolution

4.3 The governance outputs

4.3.1 Introduction

The key governance outputs of this period include the Climate Convention (UNFCCC, 1992), the Rio Declaration on Environment and Development (Rio Declaration, 1992; see, e.g. Table 4.2) and the 1995 Berlin Mandate.

4.3.2 The United Nations Framework Convention on Climate Change

The Climate Convention adopted in May 1992 (see Figure 4.1) includes a Preamble, 26 Articles and 2 Annexes. The Preamble recognizes climate change as a 'common concern of humankind'; notes that the largest historical and current GHG emissions are from the developed countries; that developing countries will need to grow to meet their legitimate priority needs; that there is need for countries to exercise their sovereign rights without causing 'damage to the environment of other states'; and promotes the use of 'no-regrets' measures that have other benefits, and context-relevant effective environmental legislation.

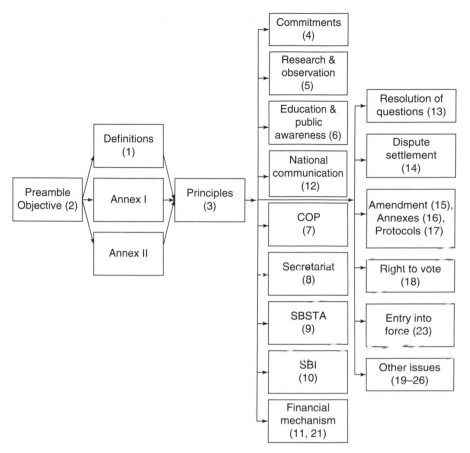

Figure 4.1   The United Nations Framework Convention on Climate Change
COP: Conference of the Parties; SBSTA: Subsidiary Body on Scientific and
Technological Advice; SBI: Subsidiary Body on Implementation
Source: Reproduced with permission from Gupta (2000), International Institute
for Sustainable Development.

The convention adopts an ultimate objective to stabilize GHG concentrations in
the atmosphere within a level and time frame that is consistent with enabling ecosystems to adapt, protect food security and promote sustainable economic development. It also divides countries into Annex I and non-Annex I countries, where
the former are the developed and the latter the developing. This was undertaken to
differentiate between the responsibilities of both groups. Subsequently, when some
Annex I countries argued that their economies were in transition (EITs) and that
they could not help developing countries, a category of rich Annex I countries was
drawn up in Annex II (see Table 8.1).

The Convention has five sets of principles which should guide in allocating
responsibility for action on climate change. These include: the principle of equity

and the common but differentiated responsibilities and respective capabilities of developed and developing countries (see 9.2.2); the need to help particularly vulnerable countries; the need for cost-effective precautionary measures; the promotion of sustainable development; and the need for a supportive open economic system. Principles are, in general, very important for countries that are less informed about a specific problem – they hope that if the principles are fair, then the allocation of responsibility will be fair. Principles are less prioritized by those (developed) countries that see them as having an open-ended character that can back-fire on them. In the negotiations, the principles were first reduced to five principles, subsequently the word 'principle' was moved from each principle to the title of the article, and finally a footnote was added that titles of the articles were only intended to assist the reader.

The convention includes commitments for Parties (countries that ratify). These include two paragraphs that set out vaguely worded targets (Bodansky, 1993: 515; Sands, 1992: 273) for the developed countries, charging them to reduce their GHG emissions to about 1990 levels by the turn of the century. All Parties need to inventorize their GHG emissions and removals by sinks (e.g. forests); make climate programmes and plans; promote education, training and public awareness; cooperate in research, monitoring, technology transfer and finance; and communicate their action to the COPs. Developed countries are required to 'provide such financial resources, including for the transfer of technology, needed by developing country Parties to meet the agreed full incremental costs of implementing measures' to prepare the national 'communications' (reports) of developing countries, and to help them with their inventories, policymaking and implementation process. The concept of incremental cost is, however, problematic (see 4.3.3). The implementation by developing countries is linked to the:

effective implementation by developed country Parties of their commitments under the Convention related to financial resources and transfer of technology and will take fully into account that economic and social development and poverty eradication are the first and overriding priorities of the developing country Parties (Art. 4.7).

The convention includes one market-based mechanism – joint implementation (Arts 3.3, 4.2, 7.2 and 11.5) – but did not define it. It was meant to imply that if the industrialized countries found it too expensive to reduce their own emissions, they could invest in projects in the developing countries to reduce the latter's emissions. This would be cheaper for the industrialized countries and would generate resources and technologies for the developing countries. In return, the industrialized countries could take the emission reductions to their side of the balance sheet. The lack of definition, and significant opposition to the idea from the developing countries, led to its eventual redefinition three years later at COP1.

The convention also allows member states of the European Communities to have joint targets. In addition, it promotes the transfer of technology to developing countries (Arts 4, 9, 11), financial transfers (Art. 11) and scientific cooperation (Art. 5).

The convention established five bodies: the COP, which meets annually to take decisions (Art. 7); a secretariat for day-to-day preparatory work and coordinating with other relevant agencies; subsidiary bodies to work on science and technology (SBSTA) and implementation (SBI); and a financial mechanism. The financial mechanism was controversial for a number of reasons. Although the idea that developed countries would need to support developing countries through financial transfers was generally accepted, the amount of such transfers was not clear. Furthermore, there was conflict regarding the location of the financial mechanism – that conflict still brews today (see 6.5). Eventually, the GEF established by UNDP, UNEP and the World Bank was given the responsibility of running the financial mechanism of the UNFCCC on an interim basis.

It also has mechanisms for reviewing national communications (Arts 4 &12), resolving implementation questions (Art. 13) and for dispute prevention, conciliation and settlement (Art. 14). All countries are expected to report on their emissions/sinks and policies (Arts 4, 12).

Although measures to deal with adaptation are rather limited, the convention refers several times to impacts, effects and vulnerability and thus implicitly prioritizes adaptation (Sands, 1992: 272). Article 3.2 focuses on the need to pay attention to the specific needs of those 'particularly vulnerable to the adverse effects of climate change' and those that 'would have to bear a disproportionate or abnormal burden under the Convention'. Article 4.4 requires developed countries to help countries particularly vulnerable to the adverse effects of climate change in meeting the costs of adaptation. Article 4.8 calls for funding, insurance and technology transfer to help specific categories of countries, including small island countries, countries with low-lying coastal areas, arid and semi-arid countries, countries prone to natural disasters, drought, desertification, high urban atmospheric pollution and fragile ecosystems, and landlocked countries. Article 4.9 focuses attention on the technology and funding needs of the least developed countries.

The convention allows for amendments by consensus (and failing that by 3/4 majority vote), annexes and protocols (Arts 15–17). The convention needed 50 ratifications to enter into force and no reservations were permitted. However, countries may withdraw three years after the convention has entered into force for them.

4.3.3 Assessing the Climate Convention

The Climate Convention sees climate change as a common concern rather than a common heritage, but recognizes that all states have a responsibility not to harm others. However, it does not elevate the 'no-harm' idea into a principle. It submits in neutral language that the developed countries have high emission levels and that the emissions of developing countries will need to grow. It also recognizes that developing countries will not be able to adopt the highest standards because of their economic situation, including the dependence of some on oil exports (i.e. the Organization of Petroleum Exporting Countries (OPEC) members).

Although the ultimate objective is phrased in qualitative and vague terms, it is given high status by being included as an article and not in the preamble (Bodansky, 1993: 497).

Only 5 principles out of the 27 listed in the Rio Declaration have been included in the Climate Convention (see Table 4.2). This inclusion is an achievement, as these were not included in the IPCC-III (IPCC-III, 1990) report. However, through a linguistic trick, they are possibly framed more as guidelines than principles (Sands, 1992: 272). It is regrettable that the no-harm, the polluter pays and/or the liability principles were not included. The precautionary principle is also somewhat weakened by the addition of the caveat that precautionary measures should be cost effective. This does not send a clear signal about the need to deal effectively with GHG emissions. Moreover, Article 3 does not explain the relationship between the principles and combines 'shall' (legally mandatory) and 'should' (not mandatory) in a confusing way (Gupta, 1997). Furthermore, while sustainable development is framed as a right and a goal (Art. 3.4), Article 4.2 argues that sustainable *economic* development is essential for adopting measures to address climate change. This article ironically focuses not on the developing countries who see themselves on the left side of the environmental Kuznets curve (see 1.5), but in an article focusing on the developed countries (Arts and Gupta, 2004). Thus, the convention accepts that development precedes sustainable development, and is thus not demanding structural change from countries. Finally, there remains enduring tension about the principle of the Common but Differentiated Responsibilities and Respective Capabilities of Countries, since some argue that it is not the per capita emissions that are important but the total emissions (e.g. Adams, 2003). This disagreement is at the heart of the continuing US refusal to adopt legally binding emission reduction measures.

The pre-1990 discussions had led many to expect that the convention would include legally binding targets for the industrialized countries, despite opposition from the USA and Turkey. However, the ultimate consensus wording might imply no clear legally binding quantitative target.

The convention prioritizes mitigation measures over adaptation. Several mitigation measures are included as well as a discussion on how to finance these. However, the adaptation section mentions insurance without elaborating on it, and ignores climate migrants (Warner and Zakieldeen, 2012) even though about

Table 4.2 *Principles in the Rio Declaration and in the UNFCCC*

Document	Rio Declaration	UNFCCC	
Principle		Preamble	Art. 3
Sovereignty subject to duty	Yes (#2)	Yes	–
Sustainable development	Yes (#3)	–	Yes (3.5)
Priority to vulnerable countries		–	Yes (3.2)
Common but differentiated responsibilities & respective capabilities	Yes (#7)	–	Yes (3.1)
Effective legislation	Yes (#11)	Yes	
Open economic system	Yes (#12)	–	Yes (3.3)
Liability & compensation	Yes (#13)	–	–
Precautionary approach	Yes (#15)	–	Yes (3.4)
Polluter pays principle	Yes (#16)	–	–
Environment impact assessment	Yes (#17)	–	–
Disaster notification	Yes (#18)	–	–
Notification of planned measures	Yes (#19)	–	–

50 million people would likely be displaced by climate change (IPCC-III, 1990: 20; Myers and Kent, 1995). Although the convention accepted that the agreed full incremental costs (Art. 4.3) of adaptation, especially for Africa (Art. 4.1 (e)) and other particularly vulnerable countries (Arts 4.4, 4.8), would be made available, adaptation financing was *not* made available. The term 'agreed full *incremental* costs' per se was supposed to imply the costs of meeting global as opposed to local benefits. As adaptation was implicitly considered a local benefit, the article could be interpreted as not applying to local adaptation. Furthermore, it was difficult to identify whether an impact was caused by GHG emissions or climate variability. The GEF clearly saw climate adaptation as leading to local benefits and did not fund adaptation in the early years. At INC10, decisions were taken to try and prioritize adaptation and, under pressure from the developing countries, COP1 decided to start identifying vulnerable countries and relevant measures. This relatively low importance paid to adaptation issues tended to ignore the IPCC-II (IPCC-II, 1996) results that changing global food production (IPCC-II, 1996: 429–30), water supplies (IPCC-II, 1996: 471–2) and vector-borne diseases (IPCC-II, 1996: 12) would affect the developing countries. A sea level rise of 50 cm to 1 metre would seriously affect small islands and their tourism income (IPCC-II, 1996: 9–10). Developing countries would suffer significantly more (2–9% of their GNP) than the developed countries (1–2%) in the event of a CO_2 doubling (IPCC-III, 1996: 183). To define adaptation therefore as a local problem leading to local benefits (Bodansky, 1993: 528) was to disconnect its link with climate change and with GHG emissions, thus trying to reduce the chances of liability discussions.

However, the convention deals with the key North–South challenges by ensuring that many southern concerns are included in its text. It ensures a close connection between the scientific world and the negotiating process through establishing the SBSTA. It encourages stakeholders to participate in the process as observers. It enhances developing-country input by allowing for developing-country negotiators to co-chair sessions and providing some resources for their negotiators to come to the meetings. It is an aspirational law (van Dijk, 1987) that seeks to promote consensus on dealing with the climate change problem, while glossing over the possible conflicts. Its strengths are its 'framework' character, which encompasses a wide range of issues and the potential for follow-up annexes, amendments and protocols; the adoption of a long-term objective and a strong normative position; the formidable institutional apparatus it created of bodies that have specific tasks; the list of follow-up tasks it generated (e.g. on reviewing the implementation process (Art. 7); the common methodology for calculating emissions and removals by sinks); the refusal to accept reservations in the convention and, thus, implicitly 'free-riders'; and the requirement of regular reports from all countries, which ensures commitment and continuity.

However, the ultimate objective is non-specific, the developed-country targets are poorly worded, the principles sometimes contradictory in nature, the division of countries into Annex I and non-Annex I is static, subjective and non-scientific, and many of the commitments are vaguely worded. From a neo-liberal perspective, the convention can be critiqued for its normative, centralized approach, for its inadequate recognition of the 'common pool' nature of the problem and the limits of international cooperation, and for the inadequate role it provides to the market.

4.3.4 The remaining INCs and the COPs

Following the adoption of the Convention in 1992, it was decided to continue the INC process to help prepare for the first COP. Among other issues, INC10 dealt with the sensitive issue of adaptation funding, proposing three stages. Stage 1 would emphasize impact studies to identify particularly vulnerable countries or regions, stage 2 would propose capacity building and other measures for the most vulnerable countries, while stage 3 would adopt '[m]easures to facilitate adequate adaptation, including insurance, and other adaptation measures' as envisaged by the convention.

This remarkably successful convention was rapidly ratified by the participating countries and has become a universal convention (see Figure 6.1). It entered into force on 21 March 1994. COP1 took place in 1995. AOSIS submitted a draft protocol text proposing that developed countries should reduce their emissions by 20% in 2005 with respect to 1990 levels. This was in line with the perspectives of some developed countries who felt the UNFCCC targets were inadequate. COP1, chaired by Angela Merkel (Germany), adopted the Berlin Mandate which established the Ad Hoc Group on the Berlin Mandate (AGBM) under the chairmanship of Raúl Estrada-Oyuela (Argentina) to promote a process that would ultimately lead to the 1997 Kyoto Protocol. COP1 also adopted a pilot phase on Activities Implemented Jointly (AIJ) to allow voluntary participation by developed-country organizations in projects in developing countries that would cost-effectively reduce GHG emissions. However, the reductions of emissions could not be credited during the pilot period to the developed countries; experiences from the pilot phase were expected to help in designing the final version of this project-based emission trading scheme. COP1 also launched the Ad Hoc Group on Article 13 to help governments deal with the challenges faced in implementation through a multilateral consultative process. This committee met six times under Patrick Széll's chairmanship (UK) before submitting its report (see Table 4.3).

The AGBM group met twice in 1995. The March 1996 meeting discussed quantitative targets and timetables, and qualitative policies and measures. COP2, chaired by Chen Chimutengwende (Zimbabwe), then *noted, not adopted* the Geneva Ministerial Declaration which called for legally binding targets for the developed countries. It also adopted 17 decisions including guidelines for developing-country national communications and on possible Quantified Emissions Limitation and Reduction Objectives (QELROS – targets) for different parties.

The Brazilian negotiator – Minister Vargas – expressed worries about the lack of 'tangible results' at COP2. Subsequently AGBM4 was requested to synthesise proposals on targets and timetables (T&Ts), and policies and measures (PAMs). In the October Bonn meeting, the USA stated that stabilization of a basket of gases by 2008–2012 was acceptable, but was subject to meaningful contribution by the key developing countries. Japan was willing to support a non-legally binding –5% target for three GHGs. In December, at AGBM5 the synthesis of proposals was discussed and the Chairman was authorized to integrate this into a 'framework compilation' – a compilation that would shape the Kyoto Protocol (see Chapter 5).

4.4 The problem definition: leadership defined

In Phase 2, the industrial emissions of the developed countries were significantly higher than those of the developing countries in both gross and per capita terms, cumulatively and annually (Bruce, 1996; Krause *et al.*, 1989: 1.5–3; SWCC, 1990: Art. 12). However, there was uncertainty about the nature of climate change and the region-specific impacts.

In this phase, climate change was defined as a global, abstract, common environmental concern of humanity. Climate change was seen as a North–South issue

Table 4.3 *COP1–2 decisions (1995–1996)*

Issue	COP1, 1995, Berlin	COP2, 1996, Geneva
Targets	The Berlin Mandate (#1)	Link between Mandate and implementation (#5)
Methods	Methodological issues (#4)	
Flexibility mechanisms	Launched Activities Implemented Jointly (AIJ) (#5)	AIJ (#8)
Finance	Interim role of GEF (#9) and its relationship with the COP (#10) including guidelines (#11), the GEF report (#12)	Guidance to GEF (#11), MOU between COP and GEF and annex on funding (#12, #13)
Other mechanisms	Technology transfer (#13)	Technology transfer (#7)
National Communications (NC)	Review of (#2) and preparations of first NCs of Annex I Parties (#3), Communications from non-Annex I parties (#8), implementation report (#7)	Guidelines for Annex I (#9) and non-Annex I Communications (#10)
Subsidiary bodies	Subsidiary bodies (#6)	Programme for SBI (#2); Secretariat technical/ financial support to Parties (#3)
Other bodies	Established AG 13 to resolve implementation questions (#20)	Future work of AG 13 (#4)
Research		Appreciate IPCC's second report (#6)
Relations with others	Link between secretariat and UN (#14)	
Budget/admin./other	(#15, #16, #17, #18, #19, #20, #21)	(#1, #14, #15, #16, #17)

and the world was divided into Annex I (North) and non-Annex I (South) countries. Although the developing countries had argued that climate change was a political issue in Phase 1, the treaty framed the problem in scientific, sectoral, technological and environmental terms, not in terms of economic or political issues (Bodansky, 1993).

The earlier discussion on liability and compensation lingered on in texts at INC4 (1994) stating that '[t]he developed countries ... should bear the primary responsibility for rectifying that damage ... and should compensate for environmental damage suffered by other countries'. Some elements remained in INC5. However, the convention text omitted all references (Verheyen, 2003).

Instead, the leadership paradigm was implicitly articulated (see Table 4.4). The concept of leadership had three dimensions – first, the leading role of the developed countries in reducing their own emissions (UNFCCC, 1992: Art 3.1). Article 4.1

calls on developed countries to take policies and measures that demonstrate that they 'are taking the lead'. Second, the developed countries were to lead by helping developing countries (UNFCCC, 1992: Arts 3.2, 4.3, 4.4, 4.5, 4.8, 11, 21) through cooperating scientifically, technologically and financially. Furthermore, Article 4.7 makes implementation by developing countries dependent on developed-country assistance. Finally, the leadership was to imply that there would be space for the economic growth of the developing countries which would possibly lead to the increase in their GHG emissions (UNFCCC, 1992: Preamble, para. 21; cf. also IPCC-III, 1990: xxvi).

The assistance to be provided to developing countries was meant to be 'new and additional financial resources' (UNFCCC, 1992: Arts 4(3), 4(4), 4(5), 4(7), 4(8), 11, 21). Article 4(3) states that developed country Parties should:

[p]rovide new and additional financial resources to meet the agreed full costs incurred by developing country Parties.

These costs are specifically in relation to emission inventories and national communications, but since the implementation of developing-country obligations was made subject to assistance from the developed countries, resources would have to be made available for such implementation as well. New and additional is to be seen as over and above official development assistance (ODA) – which was targeted at 0.7% of national income of the developed countries. However, the emphasis on the role of the developed countries in financing 'incremental costs' – the costs of global minus local benefits – presented the issue as highly abstract and distant from current issues (Gupta, 1997; Gupta and Hisschemöller, 1997).

The leadership framing included a strong normative, even idealistic, approach – an approach that recognized the differential contributions to the problem and the differential impacts of the problem. Such a framing may have arisen from the dominant role of natural scientists and engineers in the early days of the regime: a belief that environmental externalities can be internalized into the policy process; a conviction that politicizing the issue and discussing polluters and liability would not be constructive; the need to keep the issue small and concentrated; and a strong belief in fairness.

In this period, forest discussions were exported to the anticipated forest negotiations at the 1992 Rio Conference. However, no forest convention eventually materialized; only a set of forest principles. The Climate Convention did discuss forests as both sources and sinks of GHGs, and Parties are asked to conserve and enhance forests and carbon stocks. They also need to report on this in their National Communications (Haug and Gupta, 2013). The convention also exported desertification and land degradation to the Convention to Combat Desertification, and emissions from international air and water transport to other UN agencies.

Table 4.4 *Leadership articulated*

Leadership	Text
Action to reduce emissions	'The Parties should protect the climate system for the benefit of present and future generations of humankind, on the basis of equity and in accordance with their common but differentiated responsibilities and respective capabilities. Accordingly, the developed country Parties should take the lead in combating climate change and the adverse effects thereof' [Art. 3.1].
	'Policies and measures will demonstrate that developed countries are taking the lead ...' [Art. 4.1].
To help developing countries	Developed countries 'shall also provide such financial resources, including for the transfer of technology, needed by the developing country Parties ...' [Arts 4.3, 4.5; see also 3.2, 4.4, 4.5, 4.8, 11, 21].
	The extent to which developing-country Parties will effectively implement their commitments under the Convention will depend on the effective implementation by developed country Parties of their commitments under the convention related to financial resources and transfer of technology ... [Art 4.7].
To make space for growth in developing countries	The largest share of historical and current global emissions of greenhouse gas emissions has originated in developed countries, that per capita emissions in developing countries are still relatively low, and that the share of global emissions originating in developing countries will grow to meet their social and development needs [Preamble #21, cf. Art. 3.4].

Source: From Gupta (2010c). This material is reproduced with permission of John Wiley & Sons, Inc.

Thus the convention can be seen as an exercise in creative ambiguity in order to generate consensus. There was a shift from a liability framing to a leadership framing which nevertheless included many strong equity issues. However, the developed countries found it difficult to actually substantiate this in practical terms. Thus, although the two key differences between the developed and the developing countries were to be (a) the emission reduction targets of the developed, and (b) the assistance to the developing, the ambiguous targets in the convention and the lack of clarity regarding funding were problematic.

4.5 The role of actors

Following the early declarations (see Chapter 3), the world was divided into developed and developing countries. The former consisted of the USA, Japan,

Australia, New Zealand, the European Community (EC), the European Free Trade Association countries (EFTA) and some EITs. The rest of the world were the developing countries and included some of the former Second World states. This ad hoc distinction was based on the idea that the former countries had higher per capita wealth (ability to pay – capability) and GHG emissions (responsibility). However, this distinction was not based on clear and explicit criteria – a mistake that has a long legacy (see Chapter 8). The convention thus divided responsibilities accordingly – the developed countries were seen as responsible for reducing their own emissions and assisting developing countries to reduce their emissions and cope with adaptation.

The developed countries were not very well organized. While the EC had a relatively clear common policymaking framework, there were many differences between other Organisation for Economic Co-operation and Development (OECD) states, with the USA and Turkey opposing targets (see Table 3.2). The USA then engineered textual editing of the articles on targets and principles. The EITs were not very proactive. Those aspiring to EC membership supported the EC position. The EC targets of 1990 and its role were critical in promoting the adoption of the Climate Convention. While it was clear that the developed countries should reduce their emissions and help the South, some developed countries now claimed an exceptional status on the grounds that their economies had collapsed and that they were in a transition process. A separate Annex II was created in which only the richest developed countries were listed, and only these countries were now obligated to assist developing countries. This recognized the lack of ability of these countries to assist developing countries, but did not necessarily take into account their past high GHG contributions.

The developing countries consisted of about 130 G77 countries and China and some additional 20 'left-over' countries (e.g. Israel, some former East Bloc countries and some small island states). Although many of these countries had geographical, demographic, technological and financial characteristics that made them particularly vulnerable to climate change, preparing for such 'future, abstract' impacts was not prioritized in relation to current problems. Their relatively low emission levels also implied that it was not their problem. Hence, most saw climate change not as a priority, but as a mere *pseudo-agenda* item that entered the domestic agenda via the foreign policy agenda: There was very little social debate or context-related scientific information, and countries tended to make issue linkages to other issues (poverty alleviation, development, food security) and take a historical, North–South, perspective. Neither industry nor potential victims of climate change were well informed about the issue. Countries tended to move from one ideological frame to another and were at the receiving end of two Western story-lines: Liberalization (and structural adjustment programmes) and sustainable development. Links between environmentalism and traditional approaches were yet to be drawn (Balasubramanium, 1992: 33), often leading to conflicts between centralized, sometimes authoritarian, sovereign approaches and decentralized community approaches (Kuntjoro-Djakti *et al.*, 1995: E7), and between reductionist science and public interest science (Shiva and Bandopadhayay, 1986). Early discussions with developing-country actors indicated that they tended to see the future in terms of 'modernization without Westernization' as symbolizing the 'leapfrog approach' towards sustainable development. The lack of a social constituency, epistemic community and political motivation led to a dual approach – where the developing countries blamed the West for 'profligate lifestyles' but themselves aimed at Western affluence (Bhole, 1992: 295). Even in those early days, there was fear among interviewees that climate change policy might translate into a cap on growth – that it might threaten the right to develop. Within this community, two groups had a clear agenda: OPEC was single-mindedly focusing on ensuring that its dependence on fossil fuel exports was not jeopardized, and AOSIS was focusing on making its survival agenda prominent in the global arena (Gupta, 1997).

Developing countries were unified in their political perspective that climate change was a 'Western' problem and that they were the victims; this unity broke down twice in this period – once during the INC negotiations when AOSIS pushed its own agenda, and once in 1995 when the developing countries excluded the OPEC countries in forming the 'Green G77'. The Indian Ambassador drafted a text and lobbied for about one hundred signatures to support the AOSIS position for a strong Kyoto Protocol (Mwandosya, 1999). The OPEC countries were not happy about being isolated and tried to pre-empt such an occurrence in the future.

The North–South dimension of the problem although clear was not given more than lip service:

> Although industrialized countries recognized from the start the North-South dimension of the climate change issue and thus paid lip service to the interests of developing countries, the South did not forcefully express its own perspectives until later in the process.
>
> *(Bodansky, 1993: 463)*

This was possibly because the developing countries were less effective negotiators with hollow mandates (general statements of what they could say) and poor coalition-building skills. They were easily 'pacified' by offers of technology transfers and financial assistance which were not backed by measurable targets. Their unity was also often broken through the use of the word 'voluntary'. For example, pre-1990, the developing countries opposed the concept of joint implementation because it was seen as a mechanism to offset developed-country emissions and would allow them to continue emitting. However, the opposition was dealt with by the introduction of a voluntary pilot phase of AIJ in 1995 during which crediting was not allowed. The agreement of developing countries was acquired through the use of the word 'voluntary'.

Non-state actors were mostly grouped under the Climate Action Network, which took a strongly environmental and northern perspective and scarcely covered developmental issues. As time moved on, faith-based organizations were also present and a few representatives from the private sector. At the first session in 1991 there were about fifty non-state actors present, which had increased by about fivefold by 1997 (see Figure 8.8).

The Secretariat of the UNFCCC, under the able leadership of Michael Zammit Cutajar (Malta, 1991–2002) played a key role in pushing the negotiations further.

At this time, the global community of climate scientists was growing and many were engaged in the formal negotiations, either via direct participation in the IPCC or via their publications that were cited by IPCC. There was growing confidence in the scientific data on emissions, less so on the role of sinks. A considerable body of data was generated on the possible nature of impacts, and the possible solutions to the climate change problem. However, it was also made clear that equity issues, which would include the distribution of the costs of adaptation, the distribution of the costs of abatement, future emissions rights and ensuring institutional and procedural fairness (IPCC-III, 1996), could become problematic. Early controversies on the scientific predictions included a few climate sceptics who argued that: Water vapour was inadequately accounted for (Böttcher, 1992 cited in van Ulden, 1995), the role of CO_2 in stimulating plant growth was poorly taken into account (Idso, 1996), the modelling was unrealistic (Priem, 1995), and that the IPCC consensus was political rather than scientific (Emsley, 1996). However, IPCC-I (1996: 4) concluded at the time that anthropogenic causes were influencing climate.

North–South challenges to science (Gupta, 1997) included issues such as whether data on national emissions should be presented as neutral facts or that they should reflect the fact that 'survival' emissions were different from the emissions of those who lived in luxury (Agarwal and Narain, 1991, 1992) and whether emissions from cattle in the animal husbandry industry for meat and hide should be differentiated from animals for draught power and soil fertility. There were discussions that emissions from deforestation release CO_2 trapped only for some decades, while emissions from fossil fuel release CO_2 trapped for millions of years and should therefore be treated differently. A further distinction was made between justified and wrongful deforestation – where countries in the early phases of the forest transition argued that they had a right to a certain level of of deforestation. This would put the emissions of developing countries in a different light. A second issue was whether scenarios for the future had built-in biases. Parikh (1992: 508) argued that the initial IPCC scenarios assumed that income inequality would continue, allowed generous growth rates for the North and low growth rates for the South, and so suggestions for reductions by the North implied lower cuts as 'considerable fat is permitted in the reference scenario itself'. Another issue was whether the choice of 1990 as a base year adequately took past emissions into account (Krause *et al.*, 1989; Smith, 1995: 4). If past emissions had been accounted for, the permissible quota for developed countries would soon be exhausted (Agarwal and Narain, 1992). The focus on emissions related to production and not consumption also reflected a key bias. This was especially true at the time in relation to forests. Deforestation was assumed to lead to emissions in the deforesting country, while the wood was often exported to the developed countries. Furthermore, while afforestation was included in the calculations of net emissions, forest maintenance was not, and this affected developing countries. The exclusion of bunker fuels in accounting also benefitted the developed countries at the time as the bulk of the maritime and aircraft industry was in Western hands. Extrapolation of data led to exaggeration of methane and forest emissions from the South (Mitra, 1992a, b). There were also explosive discussions on the cost of human life as a follow-up to the IPCC report of 1996. IPCC-III (1996) concluded that the costs of the potential impacts of climate change to life and property in the developed countries would amount to about 1–2% of their GNP, and in developing countries about 2–9% by 2060. The reaction was in relation to whether loss of life could be valued (Douthwaite, 1995) and whether assigning different values to the lives of people based on economic wealth, gender, nationality or ethnicity was justifiable (Meyer and Cooper, 1995).

In this phase, the public was beginning to be marginally engaged especially in relation to the preparatory process of the 1992 UNCED Conference. Following the adoption of the Climate Convention, there was considerable coverage in the developed world, less so in the developing world. The Climate Convention included an article on the issue of raising public awareness in society.

4.6 Key trends in Phase 2

This was an optimistic and exciting period in climate politics. The euphoria was a coming together of two issues – the end of the Cold War and the recognition of a new problem. All the key elements of a possible climate agreement were worked out, the global community of states completely mobilized at least at national government level, and key societal actors were increasingly becoming engaged at least in the developed countries. The debate was still limited largely to the scientific and elite community and a few non-governmental organizations.

The main story line was that the leadership paradigm would provide the route towards a stable climate. The paradigm would be supported by five principles about how the world should address this problem. The paradigm would require emission reduction targets by the developed countries and assistance for the developing countries. This did not imply that developing countries should not adopt policies and measures – as laid out in Article 4. Technocratic policy options were identified by IPCC and adopted by the Climate Convention. Momentum was building up. The enthusiasm was infectious. Developed and developing countries participated and ratified the agreement at record speed. There appeared to be consensus. The problem appeared to be 'structured' in terms of the science and the values for addressing the problem.

Most developed countries were willing to adopt targets – except the USA and Turkey. This augured well for the negotiations. Although the USA helped to create confusion in the wording of the targets, the Clinton–Gore government that took over power in 1993 indicated that climate change would likely be prioritized by the US government. In the period 1990–1995, the Annex I countries had collectively reduced their emissions by 5.7% and it was expected that this would be about 3% below 1990 levels in 2000. It was anticipated that emissions would rise to 8% above 1990 levels in 2010 (FCCC/CP/1998/16/Add.1, decision 11/CP.4, Para 10 a).

However, the bulk of the emission reduction (i.e. 28%) was from the EITs, while Annex II Parties' emissions had actually grown by 3.5% between 1990 and 1995 (FCCC/CP/1998/16/Add.1, decision 11/CP.4, Para 10 b). This was not reassuring news. Underlying the consensus in the text of the Climate Convention was a growing undercurrent of resistance in both developed and developing countries. The post-Cold War euphoria and optimism was beginning to fade.

5

Progress despite challenges: towards the Kyoto Protocol and beyond (1997–2001)

5.1 Introduction

By the mid-1990s, post-Cold War euphoria was waning, most East and Central European economies were collapsing, while neo-liberal, neo-conservative ideas about rationalizing the economy were being marketed to them and the developing countries in, inter alia, the water and energy fields as a way to enhance their development potential. Sustainable development appeared elusive as a guiding concept and was being used by all to promote practically any activity.

New science confirmed that climate change was a serious problem and stabilization at 350–750 ppmv required drastic GHG emission reductions (Houghton *et al.*, 1995: 22). IPCC's Third Assessment Report (IPCC-I, 2001: 10) confirmed that 'most of the observed warming over the past 50 years is likely [66–90%] to have been due to the increase in GHG concentrations'. The recognition of the inadequacy of the United Nations Framework Convention on Climate Change (UNFCCC) targets, the Berlin Mandate, and US Vice-President Al Gore's commitment to climate change were positive signals.

However, the US Senate Byrd–Hagel Resolution (1997) signalled that the Senate would not ratify any future agreement with targets for developed countries. Japan was also being careful. The European Union (EU) hedged its bets and invested in both quantitative targets and timetables and qualitative policies and measures. The USA supported policies first and then later switched to targets.

In 1997, the Kyoto Protocol with targets, policies and a range of market mechanisms was adopted against all odds, leading the way to operationalizing its content.

5.2 The chronology of events

The IPCC 1995 report set the stage (see Table 5.1). The Ad Hoc Group on the Berlin Mandate (AGBM) met eight times to discuss targets and timetables (see 5.3).

Table 5.1 *Key events and outputs in Phase 3 of climate governance history*

Year	Activity/Outputs	Key message
1997	EU Council Conclusions	Conditional –15% target
	AGBM6: Framework compilation	Preparing quantified emission limitation and reduction targets, and policies and measures
	AGBM7: Framework compilation consolidated	
	AGBM8: Negotiating text	
	COP3, Kyoto, Kyoto Protocol	18 Decisions including Kyoto Protocol; 1 resolution
1998	White House Fact Sheet	US commitment lower than it appears
	COP4, Buenos Aires	19 decisions including: The Buenos Aires Plan of Action on strengthening the financial mechanism, technology development and transfer and AIJ; 2 resolutions
1999	COP5, Bonn	22 decisions including: guidelines for preparing National Communications by Annex I Parties, technology transfer, capacity building, flexibility mechanisms
2000	COP6, The Hague and Bonn	4 decisions and 3 resolutions; plus 2 decisions
2001		including the Bonn Agreements
2001	USA withdraws from Kyoto White House Statement	The USA decided not to ratify the Kyoto Protocol, but stays on as Party to the Climate Convention

The Ad Hoc Group on Article 13 began advising on the multilateral consultative process on non-compliance.

By March 1997, the EU (since 1993, the EC was now the EU) Council of Ministers, following long internal negotiations, was supporting a target of –15% reduction for three key GHGs for 2010 in relation to 1990 levels, on the condition that other developed countries adopted comparable targets (EU Council Conclusions, 1997). The other developed countries were reluctant to discuss commitments.

At the March 1997 AGBM, a 'framework compilation' of possible commitments was discussed and the chairman was instructed to finalize a proposal for the Kyoto Protocol. In April, Japan hosted an informal consultation. By that month's end, a negotiating text was ready and countries could submit amendments to the proposed protocol by June. In June and July, the AGBM tried to strengthen the commitments. In July 1997, US Senators Byrd and Hagel submitted the Byrd–Hagel Resolution (1997) to their Senate which stated that the USA faced high GHG emission reduction costs and should not accept binding quantitative targets unless key developing countries also participated *meaningfully* in the negotiations as:

the disparity of treatment between Annex I Parties and Developing Countries and the level of required emission reductions, could result in serious harm to the United States economy, including significant job loss, trade disadvantages, increased energy and consumer costs, or any combination thereof.

This resolution was adopted by the US Senate, leading to enhanced pressure on the developing countries to adopt voluntary commitments.

In July–August, the AGBM consolidated the commitments followed by informal meetings in September and October in Tokyo and Bonn, respectively. AGBM8 was split into two meetings with an informal consultation in Tokyo in between. Brazil proposed here a Clean Development Fund to be created from non-compliance fines levied on the developed countries who had not met their commitments under any future Kyoto Protocol. The AGBM meetings, chaired mostly by Ambassador Estrada from Argentina, prepared a consolidated text for negotiation. COP3 negotiated and adopted the Kyoto Protocol on 11 December 1997, despite objections from some countries (see 5.3).

The issue of developing-country commitments became important at the following Conference of the Parties (COPs), which also focused on operationalizing the Kyoto Protocol. 1998 was the warmest year in recorded history, probably because of strong El Niño conditions, and provided support for climate action. The incoming US president, George W. Bush, announced in 2001 (Bush, 2001) that he would not ratify the Kyoto Protocol, casting a damper on the enthusiasm of others – New Zealand, Finland, Norway and Japan (see Table 5.1). Some argue that the USA did not feel isolated, as Australia was also reluctant to ratify the protocol at the time.

5.3 The governance outputs: the Kyoto Protocol and COP decisions

5.3.1 The Kyoto Protocol

In Phase 3, the key governance output was the Kyoto Protocol and the subsequent COP decisions. The Kyoto Protocol has a preamble, 28 articles and 2 annexes (listing GHGs and developed-country party targets, respectively). It has general articles on definitions (Art. 1) and methods (Art. 5) (see Figure 5.1).

Policies and measures and other obligations

Four articles define country obligations. Article 2 discusses the policies and measures that developed-country parties should adopt in relation to their GHG emissions emitted by the energy sector, industrial processes, solvent and other product use, agriculture and the waste sector (Kyoto Protocol, 1997: Annex A). It calls for sustainable development through energy efficiency policies; the protection of sinks and reservoirs; sustainable forestry practices, afforestation and reforestation; sustainable agriculture; renewable energies, carbon sequestration

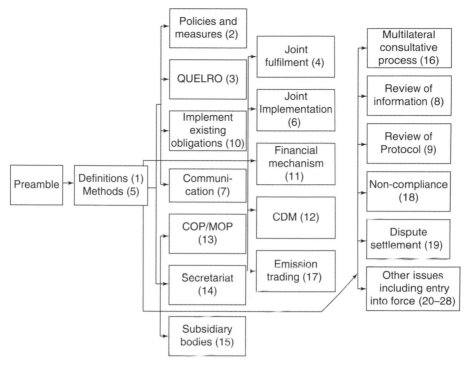

Figure 5.1   The Kyoto Protocol
Source: Reproduced with permission from Gupta (2000), International Institute for Sustainable Development.

and environmentally sound technologies; progressive phasing out of perverse subsidies; encouraging reforms in relevant sectors; controlling transport emissions; controlling methane emissions; and policies on 'bunker fuels' – fuels for the shipping and aviation sectors. Furthermore, such measures should not have adverse effects on developing-country parties – mostly the oil exporters who did not wish their exports to fall as a consequence of reduced demand from the developed world! Article 10 calls on *all* Parties, taking into account their differentiated responsibilities and circumstances, to develop cost-effective national programmes to improve the quality of local emission factors (especially as the emissions of GHGs is very contextual), to formulate and implement mitigation and adaptation measures for relevant sectors, and to report on these in their national communications. It promotes the development and transfer of environmentally sound technologies, know-how and practices; scientific cooperation; and cooperation in education, training, public awareness and capacity building. Article 7 reiterates the need for countries to improve their national inventories and develop mitigation and adaptation policies, national communications and supplementary information on country actions.

Targets and timetables

Article 3 presents the quantified emission limitation and reduction objectives (QELROs, i.e. targets) of developed countries listed in Annex B (including Annex I countries except Turkey and Belarus, and includes Croatia, Liechtenstein, Monaco and Slovenia (see Chapter 8 and Table 8.1).

The overall goal for the developed countries was to reduce their aggregate CO_2 equivalent emissions of six (not the *three* gases that negotiators were discussing previously – CO_2, CH_4 and N_2O) GHGs listed in Annex A (CO_2, CH_4, N_2O, HFC, PFC, SP6) by at least 5% in the period 2008–2012 with respect to 1990 (or thereabouts: as exceptions were made for the base year of some countries, and with respect to some gases). These countries were to show 'demonstrable progress' by 2005. These targets were with respect to *net* changes in emissions which allow the inclusion of GHG removal by human activities to enhance sinks (afforestation, reforestation and deforestation since 1990). The targets were translated into Assigned Amounts (AAs, Art. 3) of GHG emissions for five years. In order to achieve these AAs, countries could use flexibility mechanisms (see below). However, such measures should not cause adverse impacts on developing countries. Finally, where countries overachieve, the difference could be added to future commitment periods. But they were not allowed to underachieve in the hope of overachieving later – i.e. borrowing from the future. Annex B listed the individual targets of countries (see Table 5.2). Countries were allowed to achieve these targets individually or jointly; the latter took into account the EU position of having a common target.

Cooperative mechanisms

The Kyoto Protocol has four flexibility mechanisms for implementing the targets. The first mechanism – joint fulfilment – allows EU members to have a joint target (Art. 4). The other three flexibility mechanisms promote the cost-effective achievement of GHG reduction.

Figure 5.2 traces the evolution of the market mechanisms in the climate regime. Following the adoption of the concept of joint implementation in the convention, a pilot phase (Activities Implemented Jointly – AIJ; see 4.3.4) was operationalized in 1995. This AIJ should have ended with the entry into force of the Kyoto Protocol, but has been continued to allow non-parties to the protocol to invest in this mechanism. Lessons learnt from the AIJ led to the definition of project-based emissions trading in the Clean Development Mechanism (CDM) (with developing countries) and in Joint Implementation (JI) (between developed countries). It also provided the initial ideas that matured as emissions trading (ET), which entered into force in 2005. Since 2005 a forest-related mechanism has also been under discussion (see Chapters 6 and 7).

Table 5.2 *Face value of Kyoto Target unpackaged: towards dilution of the target*

Kyoto Target	Changes in Kyoto	Implications
−5.2% (in relation to 1990 levels)	From an expectation of three gases to six gases; base year for new gases may be 1995	These gases were not being emitted in 1990; they were emitted as a consequence of the phase-out of CFCs under the Montreal Protocol, despite knowledge of them being GHGs
	Inclusion of sinks	Reduces the actual effort to reduce industrial emissions; calculations become more difficult
	Different base years for EITs	Reduces the actual effort vis-à-vis 1990
	Exclusion of bunker fuels	Reduces the actual effort needed
	Inclusion of emissions trading with countries with 'hot air'	No real reduction of emissions as the reductions occur from economic depression
	Inclusion of CDM	−5.2% target includes reductions in the South; this inflates the Annex B budget

Joint Implementation grew out of the AIJ phase and allowed developed-country investors to invest in other *developed* countries (mainly countries in east and central Europe) in return for emission credits – labelled as emission reduction units (ERUs). Although Article 6 does not explicitly mention that these projects should aim at sustainable development, Article 2 which sets out the key obligations of developed countries emphasizes this. Eligible JI projects are those approved by the Parties involved (thus excluding non-Parties to the protocol); 'additional' (i.e. they should not have occurred in any case); 'supplemental' to domestic action (i.e. Parties cannot rely solely on achieving their targets through JI); and should be from Parties that have reported on national inventories of their net emissions (Arts 5 and 7).

The CDM, the 'Kyoto surprise', arose out of the reaction of the developing countries to the AIJ phase as well as their proposal for a Clean Development Fund (see 5.2). It allows entities from developed-country Parties to invest in *developing* countries in return for certified emission reductions (CERs). Such projects need to meet the criteria of sustainable development, be voluntarily undertaken, lead to 'real, measurable, and long-term benefits related to the mitigation of climate change', and the reductions should be 'additional'. As not all developing countries were happy with the offset nature of the mechanism (i.e. that emission reductions in the South could be used to compensate emission increases in the North; see 5.4) and the lack of adaptation finance, a clause was included to ensure that a percentage of the proceeds from such investments should be set aside for inclusion in an adaptation fund

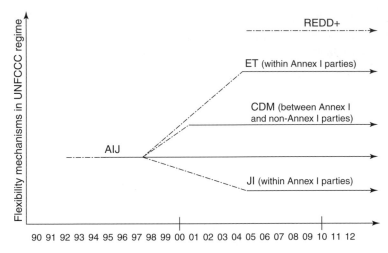

Figure 5.2   The evolution of flexibility mechanisms

(as well as administrative costs for managing the mechanism). The CDM was to be supervised by an Executive Board (Art. 12).

Emissions trading under Article 17 allowed the developed countries to trade their assigned amounts (AAs) among each other. It was included as a last-minute compromise, and became operational when the Kyoto Protocol entered into force.

Article 11 defines the financial mechanism vested on an interim basis in the Global Environment Facility (GEF). Annex II parties are required to provide adequate and predictable 'new and additional financial resources to meet the agreed full incremental costs' of preparing inventories and national communications and for providing resources for the transfer of technologies to meet 'the agreed full incremental costs of advancing the implementation of existing commitments'. In addition, the obligation to cooperate on technology transfer, science and capacity building is included in other articles.

Other articles

Three articles cover organizational issues – the COPs serving as the meeting of the Parties to the Protocol (CMP) (Art. 13), the Secretariat (Art. 14) and the Subsidiary Bodies (Art. 15). There are also articles on the Multilateral Consultative Process, Review of Information, Review of the Protocol, Non-Compliance Measures and Dispute Resolution.

A critical element is that the protocol enters into force only after at least 55 countries ratify the agreement, and those ratifying should have contributed at least 55% of the emissions of the Annex B countries in 1990. This would allow the protocol to enter into force even if one of the two (USA with 36.1% of the emissions *or* Russia with 17.4% of the emissions) did not ratify the protocol. The protocol could enter into force if the EU (24% of emissions), Russia, Japan (8.5%) and some small countries ratified, or if the EU, Russia and all other countries except the USA and Japan ratified the agreement. It was thus ensured that the entry into force of the Kyoto Protocol was not made captive to one or two large, reluctant countries. At the time, there was no information on the emission levels of Belarus, Lithuania and Ukraine.

5.3.2 Assessing the Protocol

The Kyoto Protocol has many plus points. Despite all odds, the protocol includes a target for the developed countries. It puts together a series of mechanisms to make it attractive for countries to implement the protocol without which agreement on it would have been very unlikely! Third, it further institutionalized the process and the machinery of climate implementation continued to roll on! With hindsight, I am even more positive about the Kyoto Protocol and the role of the Clinton–Gore government in pushing the international community to move further with the protocol despite the immense domestic difficulties it faced. Without the initiative of the EU in promoting strong targets, the protocol would never have emerged.

Targets and timetables

However, the very concessions that made the protocol possible are its weaknesses. The –5.2% target for the developed countries is a far cry from the –20% target suggested in the Toronto Declaration (1988) for 2005, or indeed what was needed to address the climate problem.

The targets were not 'pure'. These targets were not limited to industrial emissions but included net changes in GHG emissions by sources and removals by sinks (afforestation, reforestation and deforestation and other agreed land use, land-use change and forestry activities since 1990, but not from soils or agriculture). This gave countries flexibility to decide in which sector they wanted to take action. However, it also increased the uncertainty of the emission reductions as absorptions by sinks could not be accurately measured (Boyd *et al.*, 2008). The EU had opposed the inclusion of sinks, while the USA, Canada, Australia and New Zealand had favoured its inclusion (Gupta and Ringius, 2001; Oberthür and Ott, 1999). Second, the targets included variable base years for EITs. Third, the targets also included 'hot air' for some countries. Ukraine and Russia, whose economies had collapsed by about 30%, had significantly lower emissions than in previous years. It was expected that their emissions would remain low. They, however, successfully lobbied for stabilization targets that would allow them to bring back their emissions to 1990 levels. This was problematic as it set a precedent about acceptable emission levels. The introduction of emissions trading in combination with the high targets provided to the Ukraine and Russia implied that the excess AAs may be bought by a third party. This, of course, only leads to an exchange of financial resources, and not to any additional reduction of emissions. Europe had objected to this 'hot air'. But Russia, Ukraine and the USA thought this would be a relatively easy way to achieve financial resources during a depression for the former and target compliance for the latter. Developing countries felt cheated because they could not get targets to increase their emission entitlements which they could then sell, and also because this would imply no real emission reductions.

Another problem was that the targets included three new gases with a base year of 1995 that were not part of the problem in 1990 and should perhaps not have been emitted in the first place; however, the emission of these new gases had been promoted by decisions in the ozone layer regime. Finally, bunker fuels for marine and air transport from the high growth sector of transport were exempted from inclusion. Although the protocol entrusted the International Civil Aviation Organization with the responsibility of developing policies with respect to emissions from air transport, in the 15 years that followed no binding policies were adopted (Liu, 2011).

Together, this implied a lower target than the face value of the −5.2% target, and a delay in first steps by an additional 8–12 years (see Table 5.3). A US State Department Fact Sheet (1988) even stated that the −7% target was equivalent to 'at most a 3% real reduction below the President's initial proposal of reducing greenhouse gases to 1990 level by 2008–2012'. Furthermore, the addition of three new gases with the 1995 baseline and the inclusion of removals by sinks meant that the face value was even lower in relation to what was generally thought of at the time. For example, Australia negotiated a last-minute change to Article 3.7 which would allow it to increase its carbon budget for 1990 by including emissions from land use change and forestry (Yamin, 1998). Ambassador Slade (Slade, 1998) commented: 'And how can we be sure that these reductions will actually occur, or will industrialised countries continue to dump their waste gases into the atmosphere whilst "bubbling" their way to paper compliance'. The 5.2% reduction amounted to a marginal increase in emissions when all the 'loopholes' in the Kyoto Protocol were taken into account (Greenpeace, 1998).

Furthermore, the allocation of targets between countries suggested no real underlying criteria. The EU called for a 15% reduction of emissions; the USA was not willing to go beyond GHG emission stabilization. In the tug of war that ensued, many thought a protocol was unlikely to emerge. Eventually a compromise was reached and the EU accepted a −8% target, the USA a −7% target and Japan a −6% target. However, some countries were allowed to increase their emissions! Since EU member states could negotiate targets as high as +27% (e.g. Portugal), Norway, Australia and Iceland argued that they too could increase their emissions! Norway,

Table 5.3 *Targets of individual countries*

Party	Kyoto target*	EU member state	Pre-Kyoto willingness**	Post-Kyoto Commitment***
EU	-8%	Belgium	-10%	-7.5%
		Denmark	-25%	-21.0%
Bulgaria, Czech Republic, Estonia, Latvia, Liechtenstein, Lithuania, Monaco, Romania, Slovakia, Slovenia, Switzerland	-8%			
		Germany	-25%	-21%
USA	-7%	Greece	+30%	+25%
Canada, Japan, Poland, Hungary	-6%	Spain	+17%	+15%
Croatia	-5%	France	0%	0%
New Zealand, Ukraine, Russia	0%	Ireland	+15%	+13%
Norway	+1%	Italy	-7%	-6.5%
Australia	+8%	Luxembourg	-30%	-28%
Iceland	+10%	Netherlands	-10%	-6%
		Austria	-25%	-13%
		Portugal	+40%	+27%
		Finland	-10%	0.0%
		Sweden	+5%	+4%
		United Kingdom	-10%	-12.5%

* Kyoto Commitment for 2008–2012/1990 (& 1995 for three new gases); Base year exceptions are Bulgaria (1989), Hungary (average of 1985–1987), Poland (1988), Romania (1989)

** Council Conclusions, March 1997, for three gases

*** Council Conclusions, June 1998, for six gases

NB Belarus and Turkey had not signed the convention when the protocol was being negotiated and are not included in Annex B.

which relied heavily on hydroelectricity, did not see the potential to further reduce its emissions. Iceland's emissions were largely attributable to one aluminium factory and it wanted its exceptional circumstance taken into account.

The country targets (see column 2 in Table 5.3) reflected the position of the EU countries going into the negotiations (see column 4 in Table 5.3). During the negotiations, the EU average commitment had reduced from −15% to −8% and member states eventually took on different targets (see column 5 in Table 5.3). These individual targets were problematic as they did not set a precedent that *all* developed countries should reduce their emissions; if some of these countries were allowed to increase their emissions, a key question was in regard to why the developing countries were not allowed to negotiate corresponding increases to their emissions (which they could then trade); and furthermore, although there were clear criteria for the EU internal allocation, there were no clear criteria for the allocation of these emission levels to the Annex B countries.

Flexibility mechanisms

Let me discuss here the advantages and disadvantages of project-based emission trading at the instrument level, the organizational level and the ideological level (see Figure 5.3).

At the lowest instrument level, project-based emission trading is a cost-effective way to reduce developed-country emissions and can potentially transfer technologies to the developing countries (Forsyth, 2005). However, such projects may focus on GHG emission reduction at the cost of other sustainability issues (Mintzer, 1994), may not reflect the opportunity costs of such projects, may profit from existing subsidies on resources in the developing countries, and may exploit the cheapest options (the low-hanging fruit) leaving only the more expensive options to the developing countries. Furthermore, project-based crediting requires a baseline from which emission reductions are calculated. Determining baselines is a challenge. In some HFC 23 projects, the methodology may lead to increased emissions (Schneider, 2011). Appropriate methods can prevent such gaming (Michaelowa, 2009). Moreover, the issue of additionality – that emission reductions would not otherwise have occurred – can be checked through auditing, but that raises transaction costs. Besides, most projects do not really transfer technology (Haites *et al.*, 2006). Another issue is whether the credits should be shared by both parties and whether the export of older technologies should be accompanied by debits. Lastly the adaptation levy applied to the CDM creates a non-level field, increasing the costs of CDM projects relative to other flexibility mechanisms and, ironically, adaptation is funded by a tax on North–South cooperation.

At the middle organizational level, the project approval process is time consuming and expensive; in the early days official development assistance (ODA) was

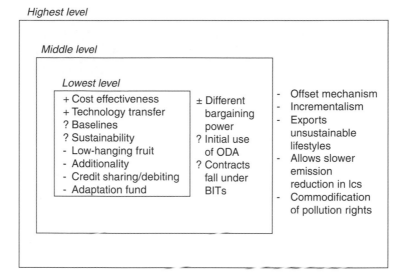

Figure 5.3   Assessing the critique on project-based emissions trading
Source: Modified from Gupta (2009). With permission of Transaction Publishers.

being diverted to fund this market mechanism; the power of the parties differs and affects contract negotiations; and the contracts drawn up are subject to complex rules of multilateral and bilateral international investment treaties (BITs) which include confidentiality of the contracts and subsequent arbitration proceedings in the event of a breach of contract.

At the highest ideological level, such projects allow for offsetting developed-country emissions (Boyd *et al.*, 2009), can lead to leakage, export unsustainable production and consumption patterns, reduce the incentive of developed countries to reduce their own emissions, and commodify pollution rights (see Figure 5.3). Despite the challenges, these market-based mechanisms have evolved over time to become successful, with the CDM delivering millions of credits.

Despite pre-1990 discussions on ET, the difficulty of allocating emission allowances to countries led to the disappearance of the topic from the agenda. ET calls for allocating a total permissible emission level among countries and countries can trade with each other. Those who overuse their budget have to buy from others, and those who underuse their budget can sell to others. This is highly efficient and ensures that all countries stay within the budget. However, as the Kyoto Protocol did not elaborate the long-term objective, there was no total budget. In dividing the budget between countries, one can use either a per capita approach or 'grandfathering'. A per capita approach would give developing countries the lion's share of the emissions (see Option 1 in Figure 5.4) and they could then sell their unused portions to the developed countries. This would force developed countries to become more

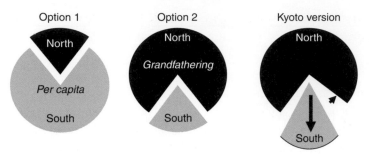

Figure 5.4 Options for emissions trading

efficient, while transferring resources to developing countries. At the same time, developing countries would need to invest in 'green' technologies in order to ensure that their own emissions do not become excessive, and could invest in adaptation. However, this would not have been acceptable to the developed countries as that would imply dividing rights worth more than one trillion dollars (Schelling, 1997).

Option 2 would require giving more emission space to the developed countries roughly in proportion to their emission levels. This is referred to as 'grandfathering'. The developing countries would get a reduced share (see Option 2 in Figure 5.4). Here the developing countries would have to buy emission rights from the developed countries in order to be able to emit GHGs while developing.

The Kyoto version of ET avoided both extremes. It allowed developing countries to increase their emissions, then allocated emissions allowances to the developed countries in line with their emissions corrected for their emission targets – some got more and some less. Together, they had to reduce by 5.2%. However, they could also buy emission reductions in the South through the CDM. This implied that the integrity of the 'developed-country budget' was not maintained, and as the developing countries were allowed to increase their emissions, the system leaks (see Kyoto version in Figure 5.4) while setting a precedent about the rules regarding allocation of emission entitlements.

The GEF

The GEF, established in 1991, was adopted as an interim entity for the financial mechanism of the 1992 Climate Convention and in the 1997 Kyoto Protocol for funding mitigation and adaptation activities in developing countries. However, the GEF funded mostly the agreed fixed incremental costs for global environmental benefits (i.e. mostly mitigation). In 1996, a Memorandum of Understanding between the GEF and the convention specified the specific responsibilities of each. In phases 2 and 3 there was much controversy about who guided the GEF and that the GEF did not invest in adaptation (Gupta, 1995). Since 1996, the GEF has funded the full costs of capacity building and research activities of Stage I adaptation activities. A few Stage II adaptation projects were also initiated. Since 1998 the GEF has been evaluated every four years from the convention perspective.

Other articles

Adaptation received only lip service in two articles (Arts 10b and 12(8)). The protocol calls on parties to facilitate adaptation and reserves a share of the proceeds from the CDM to meet the adaptation costs. The literature warned of a possible 200 million climate refugees (Myers, 2002), but the protocol did not deal with this eventuality. For a detailed analysis of the Kyoto Protocol see Depledge (2000).

5.3.3 The COP decisions

Although the protocol made considerable progress, it left implementation modalities to subsequent negotiations. The disputes between countries heightened in the coming years, however, and floundered mainly during COP6 in The Hague.

COP3 had adopted the Kyoto Protocol as one of 18 decisions. The next three COPs tried to operationalize its contents (see Table 5.4), which were ultimately finalized at the fifth meeting in Marrakesh in 2001 (see Chapter 6).

COP4 in Buenos Aires in 1998 was chaired by Maria Julia Alsogaray (Argentina). The US government put pressure on Argentina to include meaningful developing-country participation in the agenda. This caused tension between countries and ultimately developing-country opposition prevailed. COP4 eventually adopted 19 decisions which included the Buenos Aires Plan of Action – its main contribution being towards emphasizing the need to fund adaptation measures for Stage II (capacity-building) activities in particularly vulnerable countries, to enable developing countries to list their needs for environmentally sound technologies, and to provide the agreed full costs of initial and subsequent national communications. COP4 recommended simplified GEF procedures and a four-yearly review of GEF. COP4 also established a work programme on the flexible mechanisms, the terms of reference for a multilateral consultative process to help Parties deal with the problems of implementing the Convention and to prevent disputes, along with a joint working group (JWG) (not a subsidiary body) of SBTSA and SBI on compliance. The latter submitted its proposal at COP6 in 2000 under the leadership of Harald Dovland (Norway) and Tuiloma Neroni Slade (Samoa), who replaced Espen Ronneberg (Marshall Islands) after COP5. Other decisions concerned the preparations for the first CMP.

COP5 in Bonn under the chairmanship of Jan Szyszko (Poland) was unremarkable. It decided that 'Climate change was a thread in the fabric of global relations. The main breakthroughs in the climate change negotiations could be achieved only

Table 5.4 *COP3–4 decisions*

Issue	COP3, Kyoto, 1997	COP4, Buenos Aires, 1998
Targets	See KP (#1)	Impact of single projects (#16)
Adaptation	Implement UNFCCC Art. 4 (8, 9) of (#3)	Implement UNFCCC Art. 4 (8, 9) (#5)
Land use		Decisions on LULUCF (#9)
Methods	Methods for KP (#2)	
Flex. mechanisms	AIJ (#10)	AIJ (#6); WP on KP mech. (#7)
Finance	Review (#11), funding necessary (#12)	Buenos Aires Plan of Action (#1), Guidance (#2), review (#3)
Other mechanisms	Tech transfer (TOT) (#9)	Tech transfer (#4; #1)
NC	Annex I (#6)	Annex I (#11) and non-Annex I (#12)
Compliance		Joint Working Group (JWG) on Compliance and preparations for COP (#8)
Subsidiary bodies	Division between SBI/SBSTA (#13)	
Other bodies	Ad hoc group on Art. 13 (#14)	Multilateral consultative process on Art. 13 (#10)
Research	Cooperation with IPCC (#7), develop observation networks (#8)	Observation (#14)
Relations with others		Ozone Layer regime (#13)
Budget/admin./other	(#5, #15, #16, #17, #18)	(#8, #15, #17, #18, #19)
Amendments	Replaced Czechoslovakia by Czech Republic and Slovakia; and added Croatia, Liechtenstein, Monaco and Slovenia to UNFCCC Annex I (#4)	

in that global strategic context' (FCCC/CP/1999/6: para 90). It prepared the terms of reference of the Consultative Group of Experts (CGE) to help non-Annex I parties in preparing their National Communications or reports. This Group has five experts from each of the developing-country UN regions (Africa, Asia, Latin America and the Caribbean), six experts from Annex I Parties, and three experts from organizations with relevant experience meet twice annually. COP5 also promoted capacity building and focused on the special needs of LDCs. Turkey requested that its name be removed from Annexes I and II of the Convention, while Kazakhstan requested that its name be included in Annex I.

Table 5.5 *COP5–6 decisions*

Issue	COP5, Bonn, 1999	COP6, The Hague, Bonn, 2000–2001
Bunker fuel	Emissions (#18)	
Adaptation/impact	Implement 4 (8, 9), Art. 3(14) KP (#12)	
Land use	LULUCF (#16)	
Flex. mechanisms	AIJ (#13), JI, CDM, ET (#14)	
Funds		Bonn Agreement on funds and tech. transfer (#5)
Other mechanisms	Consultative process on TOT (#9), capacity building in DCs (#10), EITs (#11)	
NC	Guidance for Annex I (#3, #4), guidelines for inventory review (#6), synthesis non-Annex I (#7)/other issues (#8)	Second synthesis of Non-Annex I NCs (#3)
Compliance	Jt WG on Compliance (# 15)	
Subsidiary bodies	Consultative Group of Experts (CGE) on NCs from Non-Annex I Parties	
Research	Research/ observation (#5); cooperation with IPCC (#19)	
Relations with others	Ozone Layer regime (#17)	To UN (#6)
Budget/admin./ other	(#2, #20, #21, #22), Implementation of BA Plan of Action (#1)	(#2, #4), Implementation of BA Plan of Action (#1)

By 2000, only 33 countries had ratified the protocol with Romania as the only developed country. In order to reduce its own net emissions, the USA proposed that all 'managed lands' should be included as sinks, but this was rejected by other countries (Grubb and Yamin, 2001: 271). COP6 in the Netherlands struggled over the 150-page consolidated negotiating text. A clumsy, last-minute compromise text consisting of four sub-packages (capacity building, technology transfer and finance; financial mechanisms; land-use; and policies and measures, compliance, accounting, reporting and review) was unsuccessful, and decisions were postponed to mid-2001 (Grubb and Yamin, 2001; see Table 5.5).

Expectations were low from the resumed session of COP6 in Bonn in July 2001. Nevertheless, a compromise was achieved and three new funds to help developing countries were agreed upon. However, these political agreements were not yet translated into legal decisions.

5.4 The evolving problem definition: conditional leadership

Although the IPCC reports continued to confirm the severity of the problem, the rise of the sceptics led to questioning the climate change problem. The problem was still seen as global, but there was a greater realization that climate variability and change could affect current generations. It was perceived as challenging development, and development priorities were seen as more important than climate change. Economic research focused on the effect of unilateral or even multilateral measures on competitiveness in the developed world, and the fear that measures in the North would be rendered useless by increasing emissions in the developing world. Adaptation, capacity building and technology transfer were given lip service. However, the science in these areas was developing further. By the decade's end there was broad support for the idea that climate change and development are closely linked. There was also acceptance of the need for aid for adaptation.

Meanwhile, the nature of the leadership paradigm was changing. While in Phase 1, the core ideas of liability, compensation and leadership were explored, in Phase 2, leadership was defined in terms of reducing emissions by developed countries, helping developing countries with technology and finance, and making room for the developing countries to grow (see Table 4.4). This had made it very attractive for most developed countries that saw themselves as leaders (and not polluters) to ratify the Climate Convention. Developing countries were willing to accept that they would be followers. However, in Phase 3, there was increasing scepticism about the existence of the Environmental Kuznets curve for global GHG emissions (see 1.4.2). Some developed countries were not hopeful about being able to de-link their emissions from their growth. This was the reason behind the US Senate's position about not supporting targets if developing countries did not also accept such targets. This led the EU to announce that it would not ratify the Kyoto Protocol if the USA and Japan did not do so. And the developing countries became increasingly reluctant to ratify the Kyoto Protocol if the developed countries were not willing to show leadership as initially promised. The process had deteriorated into a conditional leadership paradigm (see 3.5.1).

Furthermore, while initially the aim was to reduce emissions *and* help developing countries, the Kyoto Protocol allowed some developed countries to increase their emissions, and the achievement of these emission targets was possible partly *via* assistance to the South. Although such use of credits was meant to be supplemental to domestic action, the COP decisions did not elaborate on what supplemental meant. While initially the convention had accepted that the developed countries should reduce their emissions in order to make space for the growth of developing countries, the Byrd–Hagel Resolution questioned the validity of allowing developing-country emissions to grow. In doing so, the issue of who was free-riding became open for discussion. Suddenly there was a focus on developing countries as free-riding on the emission reductions of the North and that northern efforts to reduce emissions would lead to large-scale leakage to the South and the loss of competitiveness of northern industry.

5.5 The role of actors

Cracks were beginning to emerge in the developed-country coalition. The EU had a fairly strong negotiating position, with its common targets and policy documents supporting this. It was also becoming clear that the member states had differences of opinion amongst themselves as to what each individually wanted to commit to, and the major policy ideas launched in the 1990s (e.g. the carbon tax) were not being implemented rapidly enough (see Chapter 8). Nevertheless, the EU was able to rally around a proposal that it would be willing to reduce emissions by 15% in 2010. Thereby, it showed leadership in the process. The USA had a democratic presidency and a very proactive environmentalist in Vice-President Al Gore when the Kyoto Protocol was being negotiated, but the Senate was not in favour of supporting far-reaching measures. In the ensuing negotiations between primarily the USA and the EU, it is difficult to see who eventually had greater influence on the design of the Kyoto Protocol. The EU had negotiated for −15%, the USA for stabilization and the end result was −8% for the EU and −7% for the USA. This would, prima facie, indicate that there was a tie in the influence. However, the ideas that led to the dilution of these targets – 'hot air' trading, the inclusion of sinks, the inclusion of three additional gases, the degree of supplementarity (the EU wanted 50%, the USA no limits) and the framing of the CDM – mostly came from the USA (Gupta and van der Grijp, 2000). Vice-President Gore (Gore, 1997) proudly claimed: 'But we hung tough, and in the end the final agreement was based on the core elements of the American proposal. And make no mistake, we stuck by the President's principles and we prevailed. The agreement reflects most of the key elements of the President's plan.' The USA then strategically decided not to submit the Kyoto Protocol for ratification and instead began to domestically implement it as far as possible. With President George W. Bush taking over at the end of 2000, the implicit support for the international negotiation process turned into explicit rejection.

Japan, as host of the Kyoto Protocol negotiations, made a brave effort to show a leadership role, despite its initial reluctance to accept targets and support only a pledge-and-review regime. Some countries that had had strong leadership positions in the pre-1990 period (e.g. Australia, Iceland and Norway) were increasingly facing domestic opposition and were now arguing against tough targets. Australia tried to negotiate a +18% target at Kyoto! All three were able to negotiate targets that would allow them to increase their emissions. For Russia, the situation was complex.

Russian scientists were still not willing to agree unequivocally that there was a climate change problem; the Russian economy and GHG emissions had collapsed; and it was rapidly losing its political influence in the world. Russia was not willing to agree to a further reduction of emissions but argued in favour of stabilizing its emissions at 1990 levels (which implied increasing its emissions back to 1990 levels by 2008–2012). In the meantime, a new group – the Central Group 11 (CG 11) – was formed to help 11 Central European states to bring their own interests forward.

The developing countries of about 130 G77 countries were increasingly becoming aware of their own differences of interests. The unity was straining in the effort to reconcile the extreme positions of the AOSIS and the OPEC countries (Yamin, 1998). OPEC had seen that the developing countries could organize as the Green G77 and took care to ensure that this would not happen again. Each regional grouping tried to focus on its own interests. GRULAC – the group of Latin American and Caribbean countries and the African Group came up with regional statements. The Asian countries remained divided. When the world was divided into Annex I and non-Annex I, only some of the former east and central European countries were invited to become members of Annex I. The others by default became part of the developing countries. But this group had few common interests with the G77 countries. At the resumed meeting of COP6 in Bonn in 2001, Armenia, Uzbekistan and Turkmenistan, on behalf of Central Asia, Caucasus and Moldova (CACAM), stated that they did not consider themselves to be developing countries. They considered themselves as countries with economies in transition (EITs).

Within the developing world, pressure from the USA led two countries to volunteer commitments for different reasons. Argentina, the host of COP8, was put under so much pressure that it volunteered to accept a commitment expressed as a dynamic target ($E = I * \sqrt{P}$) based on the relation between emissions (E measured in carbon equivalents) and GDP (P measured in Argentine pesos). I is the index value which is equivalent to 151.5. Kazakhstan, which never really felt part of the developing countries, and had seen that Ukraine and Russia had benefitted through adopting a target that would allow them to sell their 'hot air' to other developed countries, proposed that it wanted to be part of Annex I countries under the Climate Convention (but not the Kyoto Protocol).

The non-state actors were getting extremely active at this time. They were influencing the negotiation process through the reading material that they were providing to the negotiators. The large number of NGOs participating in the process tried to organize their position papers into a common statement that they would present at the negotiations. Behind the scenes, they played a most influential role in creating awareness of the politics of climate change and shaping government positions. Business NGOs were also becoming very influential. The Global Climate Coalition, established in 1989, and subsequently the Coalition for Vehicle Choice, the Climate Council and CEFIC (the federation of chemical industries) focused on finding flaws with the science and opposing measures. A body called the Carbon Club had a USD 13 million campaign against targets if the developing countries did not also have quantitative commitments, while some of the members reportedly tried to convince developing countries like China to resist targets (Oberthür and Ott, 1999: 73). Some relatively 'green' industries tried to promote a sustainable image and the implementation of the flexibility mechanisms. These included the World Business Council for Sustainable Development, the European Business Council for a Sustainable Energy Future (E^5), the European Round Table of Industrialists (ERT) with 48 captains of industry and UNICE, an employers' federation in Europe. The oil industry was divided – while some openly protested, others appeared to support climate policy (Shell, BP, Texaco). Some chemical companies claimed that they had made substantial reductions in their GHG emissions as a result of the hydro-fluorocarbon (HFC) phase-out. The car industry, afraid of being caught unawares, made a voluntary agreement with the EU to reduce its average carbon emissions. Ford and General Motors also decided to invest in energy-efficient cars and Toyota presented its hybrid car – the Prius. Insurance companies were supporting climate change by arguing that extreme weather events and their impacts were increasing. The nuclear lobby, including the European Atomic Forum, the European Nuclear Society and the International Nuclear Forum, began supporting climate change as a way to support renewed interest in nuclear energy.

The epistemic communities were significantly more organized at this time. IPCC was publishing special reports in addition to its five-yearly assessments. It published a Special Report on the *Regional Impacts of Climate Change: An Assessment of Vulnerability* (1997), looking at the specific impacts per group of countries and regions. It published a report on aviation and GHG emissions (IPCC, 1999). In 2000, it published three reports on emission scenarios; technology transfer; and land-use, land-use change and forests. These reports strengthened the basis for action on adaptation, technology transfer and forest protection, issues that were further developed in the subsequent phase of the negotiations.

5.6 Key trends in Phase 3

In this period it was becoming clear that the leadership paradigm was failing. Although the lack of targets in the Climate Convention had been compensated for by new targets in the Kyoto Protocol, these did not go far enough, the face value was higher than the actual value of these targets, and the Kyoto Protocol had not yet entered into force. The developed countries were not showing the expected leadership. Instead, conditional leadership had set in: The USA, by virtue of the acceptance of the Byrd–Hagel resolution, was unwilling to ratify the protocol until key developing countries were willing to engage in meaningful participation; and the EU had decided not to ratify until and unless the USA and Japan had ratified. Everyone was waiting for everyone else.

However, there was a silver lining: The USA had not stymied the negotiations nor walked out. In fact the USA had helped to ensure that the Kyoto Protocol included targets for all developed countries and had helped implicitly ensure that the protocol could eventually enter into force even without US participation. At home, the Democratic government did not send the Kyoto Protocol for approval to the Senate, but instead tried to implement without ratification. President Clinton announced on 22 October 1997 at the National Geographic Society that the USA would not wait till the treaty was ratified to act. Meanwhile, the EU promoted ratification by Japan and Russia. As time passed, total developed-country emissions fell by 3% in relation to 1990 levels (UNFCCC Secretariat, 1998). This resulted partly from the collapse of eastern and central European economies; the merger of East with West Germany, which led to the closure of old factories; and neo-liberal reasons for closing down the coal mining industry in the UK. At the same time, the emissions of individual countries continued to increase. The world population reached 6 billion. And many of the developing countries were taking action. 'For example, it has been estimated that China's emissions would have been 50% higher without energy reforms' (Prescott, 2000: 13). Proponents of the factor four philosophy suggested that it was possible to double wealth while halving resource use; this was essential to create space for the South to grow (cf. von Weizsäcker *et al.*, 1997).

6

The regime under challenge: leadership competition sets in (2001–2007)

6.1 Introduction

The mood worldwide was changing at the start of the new century. There was a definite movement towards supporting the neo-liberal agenda and the role of private actors while there was a growing push for lean states. Countries all over the globe began to accelerate the liberalization of their economies, and sustainable development appeared to take a back seat in this period. Furthermore, security issues became globally more important with the attack on the twin towers in New York in 2001, and high politics issues such as energy security and terrorism once more began to dominate the foreign policy agenda of countries.

However, the slow and steady administrative advances made in the Conference of the Parties (COPs) leading up to the Marrakesh Agreements breathed some life into the follow-up process to the Kyoto Protocol. The Climate Convention continued to demand policy implementation and reporting from governments. It took another five years before the first meeting of the Kyoto Protocol (CMP1) was able to adopt decisions to move the Kyoto Protocol further.

6.2 The chronology of events

As only Romania had ratified the Kyoto Protocol by 2000, President Bush's announcement in 2001 that the USA was not going to ratify the Kyoto Protocol unwittingly set into motion a fast-forwarding of action by other countries to ensure its ratification. This period led to a series of measures to implement the Kyoto Protocol. Late 2001, the Marrakesh COP7 took several decisions and prepared others for adoption by CMP1 (see 6.3.1). It established three funds and allowed a prompt start for the Clean Development Mechanism (CDM) and thereby ensured that actions were promoted in the developing world. This focus on the need for funding for environmental issues was complemented by the decision of the

International Conference on Financing for Development (2002), which reiterated that trade and investment channels should be used more coherently and that the developed countries should provide 0.7% of their national income as official development assistance (ODA).

The link between development and environment was also becoming more apparent as evident in the 2002 World Summit on Sustainable Development. However, although it produced a Johannesburg Declaration on Sustainable Development, an implementation plan, a document listing existing commitments and the Millennium Development Goals, and the promotion of 'Type II' partnerships between governments and other stakeholders, especially the private sector, it did not further build on the United Nations Conference on Environment and Development (UNCED), disappointing many in its inability to rise up to the needs of the 21st century.

Meanwhile, China became a member of the World Trade Organization and this heralded a change in geo-politics. This led to an explosion of Chinese participation in world trade and its GHG emissions accordingly grew.

The New Delhi COP8 (2002) focused attention on greater public awareness, while the Buenos Aires COP10 (2004) focused on vulnerability and adaptation (see 6.3.2). In the meantime, in 2003, Europe experienced a record heat wave providing a new incentive to push for climate relevant measures. A peak in Kyoto Protocol ratifications took place in 2004 (see Figure 6.1) including that of Russia (see 8.3.4), and the protocol entered into force in 2005. The first joint CMP took place in 2005. Unfortunately, as some countries had not ratified the Kyoto Protocol, namely the USA and Australia, there were greater administrative issues in setting up two parallel processes under the Climate Convention and the Kyoto Protocol. CMP1 adopted all previously prepared decisions on furthering the implementation of the Kyoto Protocol (see 6.3.3). That year hurricane Katrina wreaked havoc in New Orleans, also revealing chinks in US resilience to deal with climate change (see Table 6.1).

The second half of this period saw the climate change issue being redesigned as a high politics issue. The G8 Gleneagles meeting in 2005 adopted a joint communiqué and a 'plan of action' on Climate Change, Clean Energy and Sustainable Development. The influential Stern report (Stern, 2006) in the UK argued that climate change was probably the biggest example of market failure; that climate change would impact on all, but most on the poorest; that average temperatures could rise by 5°C if action was not taken; that by a 4°C rise in temperature, millions of people would be displaced and global food production affected; that atmospheric GHG concentrations should be kept to about 450–550 ppm CO_2 equivalent; and that deforestation, which accounted for about 17% of total emissions, should be tackled. The report recommended strong early action costing about 1% of global gross domestic product (GDP); this was seen as cheaper than the costs of the impacts – about 5% of GDP annually, which could rise to 20% of GDP if the non-

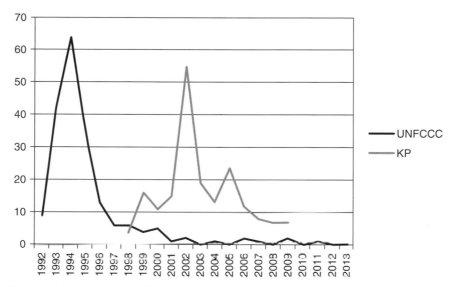

Figure 6.1   Annual ratification of the United Nations Framework Convention (UNFCCC) and the Kyoto Protocol (KP), 1992–2013, showing peaks

linear impacts eventually begin to occur. Every one tonne of CO_2 emitted caused damages of about USD 85, while emission reductions could be achieved at about USD 25 per tonne. Developing low-carbon technologies could lead to a market of about USD 500 billion. This report provided a new impulse for action in Europe. Al Gore's documentary – *An Inconvenient Truth* – also mobilized international public awareness (as well as scrutiny) on climate change.

The Intergovernmental Panel on Climate Change (IPCC) report of 2007 confirmed that climate change was a serious issue and, along with US President Al Gore, won the Nobel Peace Prize in 2007. In 2007, Arctic sea ice was melting rapidly. In April 2007, the Security Council held its first public debate on the impacts of climate change, leading to a series of discussions on the subject over the coming years. This period ended with the 2007 Bali COP12, which adopted the Bali Action Plan (see 6.3.3) to adopt post-2012 targets by COP14 in Copenhagen.

In this period the USA launched a number of competing or, as some say, complementary agreements with a number of countries to push the agenda further (see 6.4).

6.3 The governance outputs: the COP and CMP reports

6.3.1 The Marrakesh Accords, COP7, 2001

COP7, chaired by Mohamed Elyazghi (Morocco), adopted a Ministerial Declaration and the Marrakesh Accords. The declaration promotes the entry into force of the

Table 6.1 *Key events and outputs in Phase 4 of climate governance history*

Year	Activities/Outputs	Key aspects
Mid-2001	COP7, Marrakesh, Marrakesh Declaration and Accords	39 decisions: Details of measuring target achievement, flexibility mechanisms and non-compliance; prompt start CDM; Marrakesh Ministerial Declaration
2002	World Summit on Sustainable Development Declaration, Plan of Action, Type II partnerships	Need for greater private sector participation, and hybrid agreements to achieve global goals
	COP8, New Delhi	25 decisions: including a work programme on public awareness and participation; Delhi Ministerial Declaration
2003	COP9, Milan	22 decisions
2004	COP10, Buenos Aires	18 decisions, including the Buenos Aires Work Programme on Adaptation and Response Measures
2004	Russia ratifies Kyoto Protocol	Positive message to global community
2004	Kyoto Protocol enters into force	Parties have to start implementing in earnest
2005	G8 Gleneagles meeting	Decision on climate, energy and development
2005	COP11, CMP1, Montreal	15 + 37 decisions, including adopting all decisions prepared by the COP about implementing the Kyoto Protocol
2006	Stern Report	Climate change is a result of market failure, costs of action lower than inaction
2006	Al Gore's film	*An Inconvenient Truth* creates awareness and reactions
2006	COP12, CMP2 Nairobi	9 + 11 decisions
2007	UN Security Council	Debates on climate change
2007	IPCC Report published; Nobel Peace Prize with Al Gore	Reconfirms the need for action and the link between climate change and peace
2007	COP13, CMP3, Bali	14 + 11 decisions: Bali Action Plan Process for long-term cooperative action; 1 + 1 resolutions

Kyoto Protocol, and the need for addressing the adverse impacts of climate change and building capacity in the developing world. The accords adopted 39 decisions (see Figure 6.2). These included decisions on capacity building, the development and transfer of technologies, adaptation, funding, the flexibility mechanisms, implementation of their quantitative obligations (Arts 5, 7 and 8), compliance and land use.

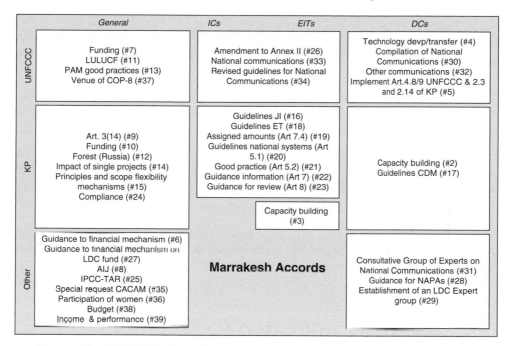

Figure 6.2   COP7: The Marrakesh Accords

Annex I Parties were encouraged to adopt and share good practices in policies and measures, and the SBSTA was to report on these to the Convention. Turkey was removed from the list of Annex II countries in the convention through an amendment subject to ratification. The accords recommended in-depth reviews of Annex-I National Communications.

A 12-member Least Developed Country (LDC) expert group on National Adaptation Plans of Action (NAPAs) (five from African LDCs, two from Asian LDCs, two from SIDS, and three from Annex II parties) was established to advise and report to the subsidiary body for implementation (SBI). The challenges facing developing countries as noted in their national communications and elaborated by the 12 member CGE were discussed and the guidelines reviewed. Two new funds were launched – the LDC fund and the Special Climate Change Fund (see 6.5).

Capacity building

The Climate Convention had not mentioned capacity building, but COPs 1 and 2 did and the Kyoto Protocol (Art. 10e) stated that Parties should:

Cooperate in and promote at the international level, and, where appropriate, using existing bodies, the development and implementation of education and training programmes, including the strengthening of national capacity building, in particular human and institutional capacities and the exchange or secondment of personnel to train experts in this field, in particular for developing countries, and facilitate at the national level public awareness and public access to information on climate change. Suitable modalities should be developed to implement these activities through the relevant bodies of the Convention taking into account Article 6 of the Convention;

Subsequent COP decisions paved the way for the adoption of a framework for capacity building for developing countries to be supported by the global environment facility (GEF), bilateral and multilateral agencies and intergovernmental organizations, to be monitored by the SBI.

Capacity building aims to 'assist them [developing countries] in promoting sustainable development while meeting the objective of the Convention' (Para. 1 of Annex to 2CP/7). The capacity building provisions point out:

> There is no 'one size-fits all' formula for capacity building. Capacity building must be country-driven, addressing the specific needs and conditions of developing countries and reflecting their national sustainable development strategies, priorities and initiatives. It is primarily to be undertaken by, and in, developing countries and in accordance with the provisions of the Convention (Para. 4 of Annex to 2 CP/7).

Capacity building is seen as a continuous, progressive and iterative process; should be effective, efficient, integrated and programmatic; should maximize synergies with other conventions; should build on existing knowledge and practices; focus on institutional capacity building; enhance and/or create an enabling environment; help in preparing national communications, national climate change programmes, GHG inventories and database management; vulnerability and adaptation assessment and measures; assessment for implementation of mitigation options; research and systematic observation; develop and transfer technology; improve decision making and assistance for the effective participation in international negotiations; the CDM; and education and training (Paras 15–17 of Annex to 2 CP/7). A similar decision (3 CP/7) was taken for capacity building in relation to economies in transition (EITs). A key concern was that capacity building would be constrained by limited GEF resources. Figure 6.3 shows the capacity building framework as envisaged. The framework would provide guidance to the GEF and this would influence implementation. A reporting and monitoring process would allow the convention bodies to review the process leading to new COP decisions (see Figure 6.3).

Technology transfer

Complaints about the lack of technology transfer and the IPCC technology transfer report (IPCC, 2000a) led COP7 to adopt a framework for transferring environmentally sound technologies to developing countries based on country needs assessment and involving relevant stakeholders. The subsidiary body for scientific and technological advice (SBTSA) was to assess these and propose assessment methods and make technology information (i.e. hardware, software and networking)

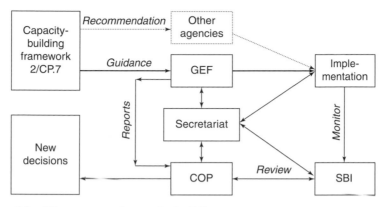

Figure 6.3   The process of capacity building

available. All Parties were to cooperate in creating an enabling environment and remove the various barriers to technology transfer, promotion of preferential procurement policies, joint research programmes and better export credit programmes. A 20-member Expert Group on Technology Transfer was established to meet biannually in conjunction with SBSTA and make an annual work programme.

Oil-exporting countries, which had continuously hampered the negotiations because of their fear that this would lead to reduced exports to the industrialized world, were now told that they would be helped to diversify their economies (see 8.4.3).

Adaptation

COP7 recommended a country-driven adaptation approach that would contribute to sustainable development, vulnerability assessment, enhancement of adaptive capacity and the transfer of adaptation technologies through the special climate change fund and the adaptation fund. SBSTA was to monitor and report on this to the COPs. LDCs were supported through a special LDC fund.

Land use, land-use change and forestry

Drawing on the IPCC's special report on land use, land-use change and forestry (LULUCF) (IPCC, 2000b), COP7 (2001) requested further work on methods, good practices, definitions and methods (11/CP.7). It prepared a draft decision for adoption at the first CMP on definitions of the eligibility of land-related activities under different articles (e.g. the management of forests, crop and grazing lands and revegetation became eligible activities for inclusion in the national targets, although there is a cap on how much land management can contribute to the target). It listed in an appendix the value of absorptions from forest management activities in the developed countries, and almost doubled the amount of absorption from forest management activities of Russia (12/CP.7).

Mechanisms

In relation to the CDM, developing-country governments could decide what is 'sustainable development' and were expected to avoid using emission reductions from nuclear facilities. Parties were urged to start such projects; not divert ODA resources for CDM; ensure reliable, transparent and conservative baselines; and promote technology transfer. COP7 established a ten-member Executive Board (one each from five UN regions, the small island developing states (SIDS), and two each from Annex I and non-Annex I Parties) to develop general and simplified rules for large and small projects, respectively. Afforestation and reforestation became eligible for inclusion under the CDM, 2% of project proceeds were to be reserved for the Adaptation Fund and administration costs, and LDC projects were exempted from this fee (17/CP.7).

On emissions trading (ET), COP7 noted that 'the Kyoto Protocol has not created or bestowed any right, title or entitlement to emissions of any kind on Parties included in Annex I', that the mechanisms aim to narrow national per capita differences, are 'supplemental' to domestic action and should promote environmental integrity (15/CP.7). This was to clarify that countries did not have rights to emit! Developed countries were also requested to avoid using emissions reductions units from nuclear facilities to meet their commitments (16/CP.7). Draft decisions on the principles, nature, scope (19/CP.7), accounting (20/CP.7) and good practices (21/CP.7) of joint implementation (JI), CDM and ET were adopted.

Guidelines for reviewing national communications (22/CP.7) included a rule that all Annex I parties had to keep a minimum of these credits as a reserve that could not be traded, to prevent the sale of too many credits.

Compliance

COP7 established a 20-member Compliance Committee with a plenary, bureau and facilitative and enforcement branches. The committee was operational following the entry into force of the protocol, would meet biannually, and decisions would require a 3/4 quorum of the members present. The committee would receive implementation questions from the secretariat, the plenary adopt policies and reports on its activities, and the bureau allocates cases to one of the committees (see Figure 6.4).

The Facilitative Branch has ten members (one from each region and SIDS and two each from Annex I and non-Annex I parties) selected for two-year terms and who can serve for two terms. It facilitates implementation by all parties and provides early warning of potential non-compliance. The Enforcement Branch has a similar composition but determines cases of non-compliance and the need for corrections and adjustments. It can suspend the party's eligibility for using the flexibility mechanisms and, where the non-complying party has exceeded its allowances,

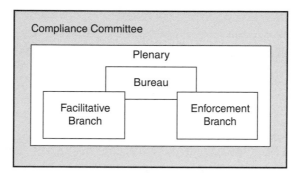

Figure 6.4   The structure of the compliance process

deduct 1.3 times the excess from the budget of the following period. Parties may appeal to the CMP (the meeting of the protocol).

6.3.2 COP8 till COP10

With so much adopted at Marrakesh, there was little to do except wait for the Kyoto Protocol to enter into force. COP8 in New Delhi, held in the same year as the World Summit on Sustainable Development, focused on climate change *and* sustainable development. Its ministerial declaration (1/CP.8) encouraged countries to ratify the Kyoto Protocol and to build on the World Summit. It adopted several decisions (see Table 6.2), and a five-year work programme on enhancing education, training, public awareness, public participation and public access to information in a holistic, cost-effective, country-driven approach. Countries and non-governmental organizations (NGOs) were requested to submit their reports to the SBSTA (11/CP.8; UNFCCC Art. 6).

COP9 in Milan adopted 22 decisions (see Table 6.2). It noted that although Annex I party emissions in 2000 were below 1990 levels, this was mostly because of reductions in EITs, and because Annex I emissions from aviation had increased by 40% in ten years, it urged for more action (1/CP.9). Milan kept the momentum up but leadership was lacking (Dessai *et al.*, 2005).

At Buenos Aires, COP10 adopted 18 decisions including the Buenos Aires Programme on Adaptation and Response measures. This programme focused attention on operational approaches to measuring, assessing and funding vulnerability and adaptation measures, and economic diversification for fossil fuel-exporting countries. It also furthered the implementation of other issues and discussed the inclusion of afforestation and reforestation activities in achieving developed-country commitments that would require differentiating between emission reduction units (ERUs), certified emission reductions (CERs), temporary certified emission reductions (tCERs), long-term certified emission reductions (lCERs), assigned amount units (AAUs) and removal units (RMUs), with simplified measures for small-scale activities.

Table 6.2 *COP8–10 decisions*

Issue	COP8, New Delhi, 2002	COP9, Milan, 2003	COP10, Buenos Aires, 2004
Mitigation (and review)	Demonstrable progress by ICs by 2005 (#25); guidelines for KP Art. 7/8 (#22)	Implementation of KP Art. 8 (#21)	
Adaptation	NAPA guidelines not necessary (#9)	NAPA guidelines not necessary (#8)	Buenos Aires Prog. (#1)
Methods for inventories		Good practice guidelines for LULUCF reports (#13), technical guidance under KP Art 5(2) (#20)	LULUCF reporting (#15) and national registry systems (#16)
Flexibility mechanisms	Continued AIJ (#14); guidance for EB of CDM (#21)	Guidance for EB of CDM (#18), on afforestation/ reforestation (#19)	AIJ (#10), CDM & TOT, certification (#12), forest treatment for ERUs, CERs, tCERs, lCERs, AAUs, RMUs (#13), simple methods for small projects (#14)
Finance	Guidance to GEF (#5) on TOT/CB (#6) on SCCF (#7), LDCF (#8)	Noted GEF adaptation ideas (#3), guidance to GEF on TOT/CB (#4) on SCCF (#5), LDCF (#6)	GEF to operationalize AF and CB (#8), need for funds (#9)
Other mechanisms (technology transfer (TOT), capacity building (CB))	New Delhi (ND) Programme on public awareness (#11); TOT (#10)	Review CB (#9)	Support ND Prog. (#7), CB in DCs (#2) and EITs (#3), TOT clearing house (#6)
NCs	DCs (#2), encouraged submissions (#4); uniform reporting format (#20); guidelines (#17/19)	Noted NCs (#1, # 2)	Standard electronic formats for KP reporting (#17)
Compliance		Code of practice for inventory review (#12)	Review of inventories (#18)
Subsidiary bodies		Request SBSTA for socio-economic analysis (#10)	

Table 6.2 (*cont.*)

Issue	COP8, New Delhi, 2002	COP9, Milan, 2003	COP10, Buenos Aires, 2004
Other bodies	Continues CGE on DC NCs (#3)	Continues LDC Expert Group (#7)	Requested LDC Expert Group to report (#4)
Research		Interaction with Global Climate Observing Systems (GCOS) (#11)	Appreciated GCOS (#5
Relations with others	Montreal Protocol (#12), desertification, wetlands, biodiversity (#13)		
Budgetary/ other	(#15, 16), ministerial declaration on climate change and sustainable development (#1)	(#22), Rules for first CMP meeting (#17)	(#11)

6.3.3 The COPs and CMPs (2005–2007)

There was much excitement as the Kyoto Protocol entered into force in 2005, and both COP11 and CMP1 were held in Montreal.

Among COP11's 14 decisions was that to enhance a non-binding dialogue on long-term cooperation on climate change that would promote long-term target formulation under the convention (see Table 6.3). The USA was opposed to such a process but, isolated with Saudi Arabia, it finally supported this provision. Another key decision was the five-year SBSTA programme on impacts, vulnerability and adaptation. COP11 requested the LDC fund provide full-cost funding for adaptation activities. It extended the mandate of the LDC Expert Group and adopted the recommendations of the Expert Group on Technology Transfer.

CMP1 was very important. It adopted the decisions prepared in previous COPs on implementing the Kyoto Protocol. It launched a process to negotiate new targets for the developed countries via the Ad Hoc Working Group on Further Commitments for Annex-I Parties under the Kyoto Protocol (AWG-KP). Its work was concluded in 2012 in the Doha Amendment (see 7.3.2) (see Table 6.3).

In this period, some developing countries argued that although afforestation and reforestation were included under the CDM, 'avoided deforestation' (i.e. maintenance of forests) was not rewarded. The Coalition for Rainforest Nations, Papua

Table 6.3 *COP11 and CMP1 decisions*

Issue	COP11, Montreal, 2005	CMP1, Longterm 2005
Targets and review	Longterm cooperation on CC (#1), enhanced review processes (#7), review of submissions of non-Annex I parties (#8), issues relating to adjustments under KP Art. 5(2) (#15)	Future commitment periods (#1) and criteria for failure of submission on GHGs (#18) Guidelines for national systems (#19), good practice guidance and adjustments (#20), adjustment issues (#21), guidelines for review # (22), terms of service for reviewers (#23), issues relating to implementing Art. 8 (#24, #25), review process for the period 2006–2007 for Annex I Parties (#26)
Adaptation/ impact	5-year programme for SBSTA on impacts, vulnerability and adaptation (#2)	
Land use	Common reporting format (#14)	Rules (#16), good practice guidelines (#17)
FM		CDM: principles, scope (#2), modalities (#3, #4, #7), afforestation and reforestation (#5), processes for small-scale forest projects (#6), of whether new HFC facilities qualify for CERs (#8); JI guidelines (#9, #10); ET: rules (#11) instructions, trading & registering (#12) and how assigned amounts could be accounted for (#13)
Finance	Guidance to financial mechanism and reporting of resource allocation framework (#6) and the LDC fund with a request to provide full cost funding for adaptation activities (#3)	Guidance to financial mechanism (#28)
Other mechanisms	Technology transfer and adoption of recommendations of the Expert Group on Technology Transfer (# 6)	Capacity building in DCs (#29), capacity building in EITs (#30), on KP Art. 3(14) (#31)
NC		Standard electronic format (#14), guidelines for reporting (#15)
Compliance		Compliance procedures (#27)
Other bodies	Continued LDC Expert Group (# 4)	Ad Hoc Working Group on Further Commitments for Annex I Parties under the Kyoto Protocol (AWG-KP) (#1)

Table 6.3 (*cont.*)

Issue	COP11, Montreal, 2005	CMP1, Longterm 2005
IPCC	Discussed research needs (#9)	
Relations with others	Links to the UN (# 11)	
Budget /admin./other	Budgets (#12, #13); flexibility for Croatia (#10)	Privileges for individuals (#33), budget (#34, #35) and other arrangements (#36), special provisions for Belarus (#32)

New Guinea and Costa Rica proposed reducing emissions from tropical deforestation (RED) through access to carbon emission markets, using national as opposed to project baselines, thus avoiding the issue of local carbon leakage. As a result, a dedicated contact group and a two-year dialogue was set up leading over time to several proposals on RED, which gradually evolved into REDD (Reducing Emissions from Deforestation and forest Degradation) and REDD+ (including all kinds of safeguards) (see Chapter 7).

COP12 in Nairobi was unremarkable, adopting nine decisions (see Table 6.4). It allowed Croatia to modify its GHG emissions in 1990 by an additional 3.5 Mt CO_2 equivalent. CMP2 incrementally improved the flexibility mechanisms, the compliance committee, the adaptation fund, capacity building and the review of the protocol. It was also decided to include Belarus in Annex B of the Kyoto Protocol.

Excited about the entry into force of the Kyoto Protocol, the Stern Report of 2006 (Stern, 2006), the Nobel Peace Prize for IPCC in 2007 (IPCC, 2007), Al Gore's mobilization of people through his documentary *An Inconvenient Truth* (2006) and Bill Clinton's Global Climate Initiative, the Bali COP13 meeting raised many expectations.

COP13 adopted 14 decisions including the Bali Action Plan (UN, 2008a; Table 6.5). This plan proposed a long-term cooperative vision, enhanced mitigation and adaptation measures, technology transfer and mobilization of financial resources. The plan called for 'deep cuts in global emissions' but did not, unfortunately, explicitly adopt a long-term quantitative target. It launched a process to develop a shared vision; long-term cooperative action; measurable mitigation actions by developed countries; Nationally Appropriate Mitigation Actions (NAMAs) by developing countries 'in the context of sustainable development, supported and enabled by technology; financing and capacity-building, in a measurable, reportable and verifiable manner'; these should lead to decisions for adoption at COP15 in Copenhagen. The Bali process was to take place within the Ad Hoc Working

Table 6.4 *COP12 and CMP2 decisions*

Issue	COP12, 2006, Nairobi	CMP2, 2006, Nairobi
Targets/review		Review of Protocol (#7)
Land use		Forest management provisions KP Art. 3(4): Italy (#8)
FM	Continuation of AIJ (#6)	CDM guidelines (#1), JI guidance (#2, #3)
Finance	Guidance to GEF (#3), Special Climate Change Fund (#1) and reviewing its operation (#2)	Adaptation fund (#5)
Other mechanisms	Capacity building (#4), extended term of the Expert Group on TOT (#5)	Capacity building (#6)
Compliance		Compliance Committee (#4)
Budget/admin./ other	Exception for Croatia (#7), administrative and financial matters (#8, #9)	Privileges for individuals serving (#9), the Belarus proposal (#10) and administrative and institutional matters (#11)

Group on Long-term, Cooperative Action (AWG-LCA), which was to meet four times before the following COP to discuss country plans submitted by early 2008.

The idea of NAMAs was one way to gradually incorporate developing countries in a flexible, 'nationally appropriate' way and was more acceptable than 'measurable, reportable and verifiable' measures. Developing countries also demanded 'measurable, reportable and verifiable' Annex II obligations on technology financing and capacity building (Müller, 2007).

The Bali Action Plan discussed risk sharing and insurance, disaster reduction strategies and other ways to deal with losses and damages resulting from climate change impacts in developing countries. However, no real link was made with compensation and/or liability (Warner and Zakieldeen, 2012).

Other decisions expanded RED to REDD (to include forest degradation) and promoted Good Practice Guidelines for LULUCF. The 19-member Expert Group on Technology Transfer under the SBSTA was reconstituted for five more years to support technology needs assessments and technology transfer, and to maintain and improve TT:CLEAR (Technology transfer clearing house). A similar, and possibly duplicate, decision was taken within the SBI. The COP also recommended that funding flows should become coherent and that the incremental cost approach be simplified.

CMP3 adopted 11 decisions (Table 6.5). Its key contribution was preparing the Adaptation Fund for operationalization in 2008 by establishing its 16-member

Table 6.5 *COP13 and CMP3 decisions*

Issue	COP13, Bali, 2007	CMP3
Targets	Bali Action Plan (#1)	Demonstration of progress by Annex I parties (#7)
Land use	Reducing emissions from deforestation (#2)	Good practice for LULUCF (#6)
Methods		
Flexibility mechanisms		CDM (#2): small-scale afforestation and reforestation (#9), JI scope and content (#3, #4)
Finance	Review GEF (#6), Guide GEF (#7)	Adaptation fund (#1)
Other cooperative mechanisms	Develop and transfer technologies under SBSTA(#3) and SBI (#4); amend New Delhi programme on awareness (#9)	
National Communications/ implementation	Compile 4th NCs (#10)	Compilation/synthesis of NCs (#8)
Compliance		Compliance (#5)
Subsidiary bodies		
Other bodies	Extend LDC expert group (#8)	
Research	Appreciate IPCC-FAR (#5), Report on Global Observing System (#11)	
Relations with others		
Budget/admin./other	(#12, #13, #14)	(#10, #11)
Amendments		

Board, a secretariat (the GEF on an interim basis), a trustee (the World Bank) and decision-making modalities. Over the years, the GEF has been controversial as many continued to argue that it did not take guidance from the COP but from its own management. Consensus was finally achieved by organizing Board meetings in Bonn rather than Washington, DC.

The AWG-KP established in 2005 now finally noted that IPCC had recommended that global GHG emissions needed to peak in the following 10–15 years and be more than halved by mid-century in order to stabilize the atmospheric concentrations at the lowest levels assessed. This would require the developed countries to reduce their emissions by 25–40% below 1990 levels by 2020. The meeting also noted that this did not take account of the fears of small island states and that it was necessary also to analyse the implications of stabilizing at below 450 ppmv of CO_2 equivalent. Parties were invited to submit their views on mitigation objectives by March 2008. The idea was that if by 2009 targets for the second commitment period could be adopted, there would be no gap between the first and second commitment period.

6.4 Climate-related agreements in other fora

With US formal withdrawal from the Kyoto Protocol, and its own perspective that the climate change problem could best be dealt with through the development and promotion of various technologies, the USA began to promote alternative agreements on climate change. Other countries also joined these agreements, either because they thought this could complement existing technology transfer approaches or because they wished to keep both the UN and non-UN avenues open as means for addressing climate change.

In 2003, the International Partnership on the Hydrogen Economy was launched which evolved into the International Partnership for Hydrogen and Fuel Cells in the Economy in December 2009. The hydrogen economy is promoted as a 'green' alternative to the hydrocarbon economy. The USA promoted this partnership, arguing that the developed countries were already investing billions in the hydrogen economy, and that such a partnership could leverage limited resources and bring together the world's best minds. The partnership included Australia, Brazil, Canada, China, France, Germany, Iceland, India, Italy, Japan, New Zealand, Norway, Republic of Korea, Russia, South Africa, the UK, the USA, the EC, the Asia-Pacific Economic Cooperative, International Energy Agency, USAID, Arctic Council and the Association of South East Asian Nations. It has promoted several projects, workshops and meetings.

In 2003, an international ministerial-level Carbon Sequestration Leadership Forum was established to promote research and development of technologies for capturing and storing carbon. Two years later the G-8 Summit endorsed this in its Gleneagles Plan of Action on Climate Change, Clean Energy and Sustainable Development.

In 2004, the USA and 13 others launched the Methane to Markets initiative which has now evolved into the Global Methane Initiative. Methane is emitted, inter alia, in the decomposition process in landfills, municipal waste and animal waste dumps, as well as from oil and natural gas production. Methane reduction has co-benefits in terms of reducing negative health impacts. If methane is recovered it can be sold as a source of fuel. This initiative promotes collaborative action to recover methane and substitute other fuels by marketing this.

In 2004 the formal structure of the Renewable Energy and Energy Efficiency Partnership was established in Bonn to catalyse the use of clean energy. Its members include Australia, Austria, Canada, the European Union, Germany, Ireland, Italy, the Netherlands, New Zealand, Norway, Spain, Switzerland, the USA and the UK, and financers such as the OPEC Fund for International Development and others.

In 2006, the Asia Pacific Partnership on Clean Development and Climate was launched. With Australia, Canada, China, India, Japan, Korea and the USA as members, the partnership aimed at promoting the use of clean energy, addressing air pollution and promoting development. It was launched with much fanfare!

These agreements can be seen as supporting the climate change treaty process; in fact the US government websites often proclaimed that that was the purpose of these side-agreements. They may have also distracted some countries from the main negotiating process, as some authors have argued.

6.5 The climate funds over time

The climate consensus was based on the idea that there would be funding for mitigation and adaptation activities for developing countries. In 1992, the GEF was named as the interim entity of the financial mechanism but tended to focus only on mitigation projects. This section discusses both the controversies surrounding the financial mechanisms (see Figure 6.5) and the evolution of the funds (see Figure 6.6). The climate convention needed a funding mechanism. The GEF established by UNEP, UNDP and the World Bank sought to unite the expertise of these three organizations in one body. And yet it has been seen as controversial over the years. While the developed countries trust the GEF because of its location within the World Bank, the developing countries were positive about the ability of the fund to deliver resources and technologies, but were more sceptical about its focus on incremental costs which implied that it did not pay for activities with local benefits. At the organizational level of analysis (middle level), the developing countries were not comfortable to knock once more on the door of the bank, there were power politics, complex procedures, and bank confidentiality rules (which have since been softened somewhat). However, the location meant that the GEF could leverage other funds and try and green the World Bank. At the highest and most abstract level of analysis, the controversy was with regard to the ideological setting of the World Bank (its neo-liberal starting point) and the definition of the sustainability problem as being rooted in the South rather than the North (see Figure 6.5). But most importantly, the UNFCCC which sought to control the GEF has engaged in a continuing battle over the years. For long, the GEF was seen as

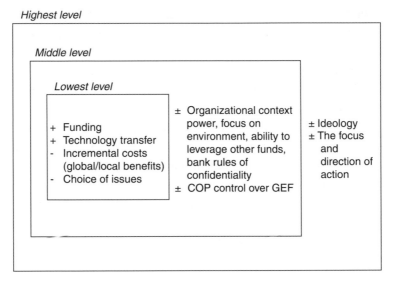

Figure 6.5   The North–South challenges to the GEF

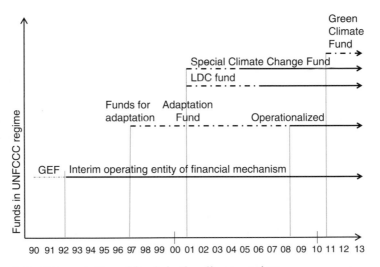

Figure 6.6   The evolution of funds in the climate regime

the interim funding mechanism for all the funds and these controversies lasted till Copenhagen (see 7.3.2) and beyond.

By 1997, its position was reaffirmed but the COP sought control over the GEF. A dedicated fund for adaptation was launched in 1997, concretized in 2001 and went into operation in 2008. Two other funds were launched in 2001 – the LDC fund and the Special Climate Change Fund. These funds were all entrusted to the GEF. In 2011, a new Green Climate Fund was launched (see Chapter 7).

The Special Climate Change Fund was aimed at financing adaptation, transfer of technologies in the area of energy transport, industry, agriculture, forestry and waste management and to help developing countries diversify their economies. On adaptation, it was to help with the continuation of preparation and/or completion of initial National Communications to the Convention (Stage I activities) and strengthen the implementation of country-driven adaptation activities that build upon the work done at the national level, either in the context of National Communications to the Convention or in-depth national studies (Stage II activities). It was to support the 'country team' approach, regional information networks, data collection, climate change-related institutions and centres of excellence, and help with national communications, public awareness, disaster preparedness activities, and early warning systems for extreme weather events. The projects needed to be country-driven, based on national priorities and aimed at sustainable development.

The LDC Fund, implemented through UNDP-GEF, supports LDCs by meeting the full costs of preparing their NAPAs (lasting about 12–18 months and for a maximum of USD 200,000 each). The NAPAs should identify priority areas for climate adaptation in LDCs including capacity building and priority projects through a participatory and multi-disciplinary country-driven process. The criteria for determining priorities include the level or degree of adverse effects, poverty reduction to enhance adaptive capacity, synergy with other conventions and cost effectiveness, potential loss of life and livelihood, human health, food security and agriculture, water availability, quality and accessibility, essential infrastructure, cultural heritage, biological diversity, land-use management and forestry, environmental amenities and coastal zones (28/CP.7). However, the link between the NAPAs and the new and additional financial obligations of the developed countries remained unclear (Verheyen, 2003).

The Adaptation Fund was to finance concrete adaptation projects including the avoidance of deforestation, combating land degradation and desertification. (This decision was a reaction to the argument that only those who deforested were receiving assistance to afforest, but those who were maintaining their forests were not – i.e. the deforesters were being rewarded.) The funds would be raised from a 2% share of the proceeds from the CDM and other sources of funding. The Fund is managed by the CDM Executive Board and receives guidance from the CMP. What share and type of adaptation projects was to be financed in which countries under the Adaptation Fund was not yet agreed.

In follow-up, the Marrakesh Accords decided that the GEF would remain the interim entity of all the financial instruments. But the conflict over control over GEF activities continues to date, and practically every COP has tried to control the GEF and call for simplifying procedures.

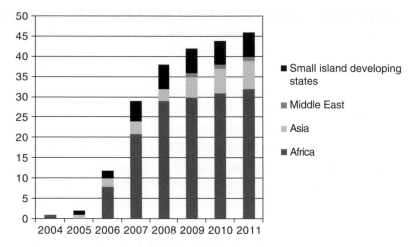

Figure 6.7   Cumulative totals of NAPAs
Source: Derived, with permission, from the database of Roger *et al.* (2013).

6.6 The problem definition

Reviewing the developments in Phase 4 of the negotiations, climate change continued to be seen as a common concern but its links with sustainable development became evident. 'Climate change is the defining development issue of our generation' (IPCC-II, 2007; IPCC-III, 2007; UNDP, 2007).

The leadership paradigm was flagging in this period. The EU was pushing for action. Many of the developed country Parties were working hard to meet their Kyoto commitments. However, the US and Australian decision not to ratify the protocol had led to considerable pessimism. Nevertheless, the persistent development of the COPs and EU pressure on countries to promote ratifications led to the ratifications reaching their highest point in 2004; the protocol entered into force in 2005 (see Figure 6.1). In December 2007 following a drought and change in government, the Australians finally ratified the protocol thereby isolating the US position. During this period, the USA began to engage in leadership competition by promoting a number of different instruments that directly or indirectly dealt with climate change (see 6.4).

The LDC Fund, specially established to assist the poorest countries with their National Adaptation Programs of Action (NAPAs) and implemented through UNDP-GEF, aimed to finance the full costs of preparing NAPAs, up to USD 200,000 per country, and these NAPAs were to identify priority areas addressing the urgent and immediate needs and concerns for adaptation to the adverse impacts of climate change in LDCs. This led to a spurt of NAPAs (see Figure 6.7).

As climate became increasingly linked to development, it also became linked to development cooperation and there was considerable talk of mainstreaming climate change into development cooperation. In the meanwhile, it was also becoming increasingly clear that the developed countries did not have significant new and additional resources to make available. Most were unable to make 0.7% of their national income available for development assistance. New and additional was now being interpreted as the new and additional to *existing* development assistance. Furthermore, the discussion was shifting towards mainstreaming climate change in development assistance. Donor agencies, the European Union, the G7, the World Bank, the UN Environment Programme (UNEP) and others began to link climate change to development assistance and the OECD-DAC adopted a Declaration on Integrating Climate Change Adaptation into Development Cooperation in 2006 (see Table 3.1 of Gupta, 2010a: 69). Arguments in favour of mainstreaming climate change into development assistance included: scientists were favouring an integrated (not sectoral) approach; aid agencies were fighting aid fatigue and seeking new arguments for funding and wishing to deal with past critique on the environmental impacts of aid projects; the climate lobby groups were seeking resources and, given the few resources available, were trying to promote cost-effective spending; developed countries felt that mainstreaming would reduce the need to generate new and additional resources; and multilateral agencies argued that the link would enhance efficiencies (see Table 3.2 of Gupta, 2010a: 72).

At the same time, there was a clear need for extra resources. The hope was that the market would generate these resources. US non-participation in the Kyoto Protocol had led to reduced demand for emission reduction credits leading to a fall in the price of carbon. Nevertheless, the CDM was becoming successful and mobilizing thousands of projects with about a thousand million registered CERs. This helped developed countries achieve their targets. The EU's own ETS was being operationalized.

Another critical issue was the way forests were being dealt with. Afforestation and reforestation projects were included in the CDM, but credits generated had an expiry date and could be used only up to 1% of the Party's base year emissions in the first commitment period. However, there was much discussion on how to deal with deforestation and, conversely, how to reward those countries that were protecting their forests. While the EU and some small island states saw inclusion of 'avoided deforestation' as reducing the impetus for reducing emissions, others saw it as encouraging forest maintenance. By 2005, this issue was on the international agenda.

Meanwhile, the search for alternative fuels focused attention on biofuels. Many countries, including those of the EU, developed policies on promoting and importing biofuels leading to a spurt in global biofuel production. However, this increased at the cost of food security, environmental pollution, high water use and new social problems.

There was also increasing fear among scholars and NGOs alike confirming Myers' earlier assertion that the climate problem would give rise to some 150–200 million refugees by 2050 (Brown, 2008: 11; Jakobeit and Methmann, 2007).

6.7 The role of actors

In this period, Lomborg (2001: 326) had claimed that environmentalists and scientists were prophets of doom, emphasizing the modern 'dysfunctional civilization' while ignoring the progress made in tackling security, access to basic needs, health, education and scientific knowledge. He argued that environmental problems were less urgent than the developmental challenges, that society overestimates small risks and underestimates big ones, and that global resources should be spent based on prioritizing global problems.

However, IPCC-I (2007) continued to warn of looming problems. GHG concentrations had increased from 280 to 379 ppm in 2005 and the annual rise in sea levels around 1961 (1.8 mm) had increased to 3.1 mm in recent years. If average temperatures were to be kept below 2°C, GHG concentrations would have to be stabilized at 450–490 ppm with peaking about 2015. IPCC-III (2007) anticipated in particular that the energy sector could increase its emissions by 40–110% more in 2030 with respect to 2000. But it also argued that several measures could help countries reduce their GHGs by 6 GT CO_2-eq/year by 2030. But this called for political will. Stern (Stern, 2006) argued that if no further action was taken, countries would face risks 'equivalent to losing 5% of global GDP each year, now and forever'. If other risks were included, the costs would be higher.

At this time, a series of 'hockey stick' graphs showed that emissions had been quite stable for 10,000 years and then suddenly rose dramatically. Although the scientific validity of the graphs was questioned, IPCC (2007) reiterated through the use of other sources of data that the shape was correct. Thus, although the IPCC documents were reiterating earlier findings, while nuancing them, there was increasing scepticism on the climate change problem.

In the social sciences, scholarly attention focused on the post-Kyoto market mechanisms and on burden sharing to help address the climate change problem and on mechanisms of engaging local government (Cao, 2010; Chakravarty *et al.*, 2009; Den Elzen and Höhne, 2010; Frankel, 2010; Mearns and Norton, 2010).

In terms of coalitions, the EU was growing – from 25 countries in 2004 to 27 in 2007 – and becoming institutionally stronger. The Umbrella Group that had united some of the non-EU Annex I countries was falling apart, with the USA and Australia not ratifying the Kyoto Protocol. The Central Asia, Caucasus and Moldova (CACAM) group was desperately trying to find a voice in the process. The small island states were becoming more vociferous but differences were beginning to show between countries. OPEC continued to obstruct measures. Although the G77 operated as one group, internal differences became more and more obvious. The LDCs were beginning to get a grip on the adaptation risks that they would be facing, as each year the number of incoming NAPAs increased (see Figure 6.7 and Chapter 8).

The situations within countries were diversifying. Sub-national authorities in the USA, Australia and Canada were becoming active, possibly because national-level commitment was low. For example, New York State, Connecticut, Maine, Massachusetts, New Hampshire, Rhode Island, Vermont, Newfoundland, New Brunswick, Nova Scotia, Prince Edward Island and California were all developing their own climate policy. Sometimes such action was being taken because of the vacuum at national level, and sometimes because authority had been delegated through processes of decentralization to lower authorities as in France, Italy and the Netherlands. Not only were local authorities being mobilized, but courts in different countries were also having to deal with climate-relevant cases (see Chapter 9).

While environmental NGOs lobbied heavily on climate change, new actors were becoming important. Prominent figures such as Al Gore and Bill Clinton were pushing for climate change policy, each through their own initiatives. Religious organizations were espousing the climate change issue. Women's groups began to focus on the link with women. Industry was split into different groups – some resisting action, others supporting it.

The climate secretariat was playing a key role in promoting action and was led at this time by Joke Waller Hunter (2001–2005) and Yvo de Boer (2005–2010), both from the Netherlands.

6.8 Key trends in Phase 4

In the fourth phase there was flagging leadership from the North and the promotion of alternative ways of addressing the problem. The flagging leadership was not just evident in delayed ratification (if at all) – it was evident in the way various paragraphs of the agreements were being implemented. For example, Article 2 of the convention on dangerous climate change was quantitatively articulated only in 2007. Such articulation was necessary because it was important to know what the threshold levels should be in order to ensure that current actions were in line with these levels (Gupta and van Asselt, 2006; Hare and Meinshausen, 2004; Oppenheimer and Petsonk, 2003; Ott *et al.*, 2004: 27). Previously, G.W. Bush (2001) clearly stated: 'No one can say with any certainty what constitutes a dangerous level of warming, and therefore what level should be avoided'. Many were arguing that the level at which climate change becomes dangerous could not be determined scientifically as it was a political question (Brooks *et al.*, 2005; Dessai *et al.*,

2003; Gupta *et al.*, 2003; IPCC, 2003, 2004; O'Neill and Oppenheimer, 2002; Oppenheimer and Petsonk, 2003, 2005; Ott *et al.*, 2004). The EU had decided that temperatures beyond 2°C above pre-industrial levels were likely to be dangerous. The flagging leadership was also evident in retraction from the commitment towards new and additional resources through the discourse on mainstreaming climate change into development cooperation. It was evident in the search for new ways to deal with climate change leading to a flurry of agreements on hydrogen, methane, renewable energy, carbon capture, biofuels and clean development. A fact sheet of the USA in 2004 revealed that all these side-activities were to be seen as actively contributing to addressing the climate change problem (White House, 2004). On the positive side, the EU was becoming stronger and it remained committed to dealing with the climate change problem through a multilateral approach. The National Communications from Annex I Parties revealed that GHG emissions had fallen by 6% over 1990–2008, resulting from the restructuring in the EITs, policy implementation including with respect to renewables and energy efficiency and the economic crises (UNFCCC Compilation, 2011).

The reluctance of some governments to take action was compensated by lower government agencies and the judiciary (see Chapter 9) becoming more active in dealing with these issues. The market mechanisms were becoming effective.

7

Enlarging the negotiating pie (2008–2012)

7.1 Introduction

The previous period had ended with IPCC (2007: 30) arguing that: 'warming of the climate system is unequivocal', and with IPCC and Al Gore winning the Nobel Peace Prize. Several high-level global climate meetings were held including those at the UN Security Council and the G8 summits. With US President Barack Obama coming to office there was an expectation that climate change was going to be given serious attention.

However, this was also the beginning of an explosion of climate scepticism. Furthermore, by 2008, the global recession was setting in – affecting the developed countries while the large developing countries appeared to have booming economies. China, South Africa, Brazil and India were becoming richer. The international dynamics were changing. The question was whether the developing countries would follow the developed countries into their spiralling recession and thereby have significantly lower emission increases, or whether they would remain relatively unhurt.

7.2 The chronology of events

In the run up to 2008, there was a housing crisis (prices first went up and then fell) followed by a financial crisis in the USA. This spilled over to Europe spurred by investments in subprime mortgages in the USA, collapsing banks and lenders (UK Northern Rock, US Bear Stearns, Lehman Brothers, Fannie Mae and Freddie Mac), leading to financial problems in other banks and the start of a recession with falling growth rates in many countries. Some countries began to recover, but Greece, Iceland, Ireland, Portugal and Spain continue to face problems. The European Central Bank adopted several measures. Recovery has slowed down in most developed countries. The developing countries appeared to be growing but exports will probably fall as the recession continues.

In 2008 oil prices reached a high at USD 147 per barrel, providing an impetus to search for other sources of energy including shale energy. Shale energy seems to promise energy independence to countries like the USA, possibly locking the USA into a fossil fuel-dependent future. Meanwhile, Cyclone Nargis devastated a part of Myanmar. In 2010, one fifth of Pakistan was flooded. All these events raise public consciousness about climate change. In 2011, an earthquake and tsunami killed over 9000 people and cost about USD 300 billion and shook faith in nuclear power in Japan. By the end of the year, the global population had reached the 7 billion mark. The recession and public scepticism reduced the impetus for action in the Conference of the Parties (COPs) (see Table 7.1).

However, in 2012 the world's first 1-gigawatt offshore wind farm was initiated in UK waters aiming to reduce carbon emissions by 2 Mt per year in order to meet its own target of reducing emissions by 2050. Germany decided to phase out nuclear power and promoted decentralized renewable energy. Sub-national authorities began to initiate climate-relevant measures. Fear of global inaction led simultaneously to action on human rights. In 2011, the Cochabamba conference called for reducing GHG emissions by 50% in 2017, that developed countries should acknowledge their climate debt, recognize the human rights of people including indigenous peoples, establish an International Court of Climate Justice, reject the commodification of nature, and suggested that 6% of GDP should be paid by the developed countries towards climate compensation (People's Agreement, 2010; see also Chapter 9).

7.3 The COPs and CMPs

7.3.1 COP14–15, from Poznan to Copenhagen

At COP11 (2005) in Montreal, a twin track had been set into motion – one involving the Climate Convention parties on long-term, cooperative action institutionalized in COP13 (2007) in Bali (AWG-LCA), and one involving the Kyoto Protocol Parties (AWG-KP). One of these tracks would hopefully lead to new quantitative commitments for the developed countries: COP14 (2008) in Poznan was the halfway halt. However, the EU was still struggling internally with its own target (effort) sharing negotiations and the USA was waiting for a new president.

The AWG-LCA discussed monitoring, reportable and verifiable (MRV) commitments for the developed countries, nationally appropriate mitigation actions (NAMAs) for developing countries, and reducing emissions from deforestation and forest degradation (REDD); adaptation issues; technology development and transfer; and financing and investment. The six sessions of the AWG-KP held since Montreal had discussed the mitigation potential and possible objectives for the

Table 7.1 *Key events and outputs in Phase 5 of climate governance history*

Year	Activity/Event	Key messages
2008	Recession	Reduces impetus to take action, but emissions fall
2008	COP14, CMP4	9 Decisions: Launched adaptation fund and Poznan Technology mechanism; 8 Decisions
2009	US president Obama comes to power	Expectation of hope that there will be a reversal in US climate policy
2009	COP15, CMP5	13 Decisions: including Copenhagen Accord, notes long-term objective; subsequently – countries submit voluntary commitments; aim to raise USD 30 billion in 2012 rising to USD 100 billion in 2020; 10 Decisions
2010	COP16, CMP6	12 Decisions: Including the Cancun Agreements and establishing the Green Climate Fund; 13 Decisions
2010	Recovery starts?	Reduced impetus to take action, but emissions rise
2011	Tsunami in Japan	Reduces faith in nuclear options as a way to reduce GHG emissions
2011	COP17, CMP7	5 Decisions: Process for targets for developed and developing countries to be ready by 2015 and binding from 2020 (Durban Platform); launch of Green Climate Fund; 17 decisions
2012	Recovery slows down Hurricane Sandy	Shows how vulnerable relatively rich parts of the USA can be to extreme weather events, leading to some focus on climate change
2012	COP18, CMP8	26 Decisions 13 Decisions: Including amendment on second commitment period (KP) – 2012–2020; process for adopting agreement in 2015 and binding from 2020

Kyoto Protocol Annex B Parties. At Poznan they discussed the need to reduce emissions by 25–40% in 2020/1990 as recommended by IPCC-III (2007). The EU led by suggesting that it would reduce its emissions by 20% by 2020. Australia, Brazil and Mexico also made pledges. Others were reluctant to make pledges till the US president was elected. Having two parallel processes discussing similar issues is inefficient, and New Zealand proposed merging the processes. Developing countries vetoed this because the goals and the parties of the two processes were distinct. 'In their words, the AWG-KP was the channel "to show leadership"' (Santarius *et al.*, 2009: 78).

At Poznan, adaptation was finally given some importance. Ten years after the idea had been proposed, the CMP operationalized the Adaptation Fund (see Figure 6.6) and established the Adaptation Fund Board under the COP with the global environment facility (GEF) serving as the interim secretariat (1/CMP.4).

This was a small victory for the developing countries that had tried since the early 1990s to ensure that the fund was not run too independently of the COP (see 6.5). However, although the Adaptation Fund Board did get 'legal capacity' this was only granted 'if necessary' (Garnaud, 2009). The developing countries felt that the GEF services were too expensive for a small fund, and that the World Bank, as Trustee of the Fund, had a conflict of interest as it both monetizes the certified emission reductions (CERs) and buys and sells them. The World Bank retorted that different parts of the bank did different things (Santarius *et al.*, 2009). The Adaptation Fund was to be primarily funded by a 2% levy on CDM projects; in August 2008, 3.2 million CERs were held in the Adaptation Fund CDM account; the money this raised was relatively small compared with the adaptation needs (Santarius *et al.*, 2009). Extending the 2% levy to the other flexibility mechanisms was suggested once more by Costa Rica and South Africa, while the developed countries argued against disturbing these market instruments. Several suggestions were made for creative financing: Norway suggested that 2% of the permits under the assigned amount units (AAUs) could be auctioned to generate funds; Mexico a Multilateral Climate Change Fund or a Green Fund; and Switzerland a global CO_2 levy of USD 2 per ton on fossil fuel emissions. The G77 and China proposed that Annex I Parties should give 0.5–1% of GNP for the financial mechanisms (Santarius *et al.*, 2009). Insurance issues were also given greater attention.

Issues inadequately discussed included the criteria for allocating funding, which parts of adaptation projects were eligible for funding and whether the money for adaptation be compensation or assistance:

> Is it assistance, in the same vein as ODA, or is it compensation on the basis of the polluter pays principle? From the developing world's point of view, there is no doubt that it is not ODA; and in the developed countries' minds it should surely not be compensation as they do not want to be held officially (and financially) responsible for all the damages caused by climate change.
>
> *(Garnaud, 2009: 3)*

COP14 also tried to promote the development and transfer of technologies through a Strategic Programme on Technology Transfer for developing countries to be funded by the GEF (2/CP.14) (Lovett *et al.*, 2009).

Other COP decisions were mostly incremental decisions in advancing the Bali Action Plan (see Table 7.2). Discussions on REDD resulted in expanding its scope to include 'the role of conservation, sustainable management of forests and enhancement of carbon stocks in developing countries' (UN, 2008b: 1). The Poznan COP was unremarkable except for showing some degree of political determination and leadership (Santarius *et al.*, 2009).

Expectations were were very high for Copenhagen. New quantitative commitments were expected. Barack Obama had just become US president. The EU had

Table 7.2 *COP14–15 and CMP4–5 decisions*

Issue	COP14, Poznan, 2008	CMP4	COP15, Copenhagen, 2009	CMP5
Mitigation targets	Advancing Bali Action Plan (#1)		Copenhagen Accord (#2)	
Methods			Guidance for REDD and carbon stock (#4)	
Flexibility mechanisms	AIJ (#7)	CDM (#2) JI (#5) Guidance		CDM (#2), JI (#3) Guidance
Finance	GEF review (#3) Guidance (#4), and LDCF (#5)	AF (#1)	GEF review (#6), guidance (#7)	AF Board Report (#4), AF Review (#5)
Other mechanisms	CB (#6), TOT (#2)	CB (#6)	CB (#8)	CB (#7)
Review and compliance		Compliance Committee (#4)	Training for review experts (#10)	Training for review teams (#8), Compliance Committee (#6)
Other bodies		AWG-KP (#3)	Outcome of AWG-LCA (#1); of CGE of Experts on non-Annex I NCs (#5); High Level Panel on Green Climate Fund	Outcome of AWG-KP (#1)
Research			Climate observations (#9)	
Budget/admin./ other	(#8,#9)	(#8), Privileges/ immunities for individuals (#7)	(#11, 12, 13)	(#8)
Amendments			to Annex I (#3)	

prepared both an unconditional (a 20% reduction in 2020/1990 levels) and a conditional target (a 30% reduction in 2020/1990 levels if other developed countries adopted measures). Most developed countries had something to offer: Norway was willing to reduce emissions by 40% and Japan by 25%. The USA offered −17%

from 2000 levels (which amounts to −4% from 1990 levels). Many developing countries, for the first time, too, had commitments to offer: Brazil, China, India, Indonesia, Mexico and South Korea, among others. Denmark was a proactive host. Many heads of state were present. The omens were positive.

COP15 took 13 decisions (see Table 7.3) including the Copenhagen Accord. This Accord agreed for the first time on an explicit long-term goal recognizing the need to keep global temperatures below a 2°C rise from pre-industrial levels, and to also look into the possibility of a 1.5°C target. It requested all countries to adopt emission targets. It decided on the Copenhagen Green Climate Fund with USD 30 billion annually in 2010–2012 rising to USD 100 billion annually in 2030 to be promoted by a High Level Panel; and a technology mechanism. This accord, however, was only *noted* by the parties as only 122[1] rising to 139[2] countries agreed to it.

Some developing countries did not accept the accord because, although it included a quantitative long-term target, countries were allowed to come up voluntarily with their own targets by January 2010. While the funding decision was welcomed, it was not clear who would raise the money, by how much and what would happen if they did not do so. Although the funding decision was in line with ideas proposed by Meles Zenawi (the late Prime Minster of Ethiopia), it fell short of what he had proposed in terms of financing, 'fast-start', priority financing and a High-Level Advisory Group on Climate Change Financing (Roger and Belliethathan, unpublished).

In January 2010, many countries submitted their pledges (see Table 7.3). While most industrialized countries had 1990 or thereabouts as a base year, Australia, and USA and Canada chose 2000 and 2005, respectively. The EU-27, Liechtenstein, Kazakhstan, Switzerland and Australia adopted unconditional and conditional targets. The rest adopted only conditional targets. These were conditional on global commitment towards a long-term target, similar commitments from other developed countries, and commitments from developing countries. Belarus, which

---

[1] Albania, Algeria, Armenia, Australia, Austria, Bahamas, Bangladesh, Belarus, Belgium, Benin, Bhutan, Bosnia and Herzegovina, Botswana, Brazil, Bulgaria, Burkina Faso, Cambodia, Canada, Central African Republic, Chile, China, Colombia, Congo, Costa Rica, Côte d'Ivoire, Croatia, Cyprus, Czech Republic, Democratic Republic of Congo, Denmark, Djibouti, Eritrea, Estonia, Ethiopia, European Union, Fiji, Finland, France, Gabon, Georgia, Germany, Ghana, Greece, Guatemala, Guinea, Guyana, Hungary, Iceland, India, Indonesia, Ireland, Israel, Italy, Japan, Jordan, Kazakhstan, Kiribati, Lao People's Democratic Republic, Latvia, Lesotho, Liechtenstein, Lithuania, Luxembourg, Madagascar, Malawi, Maldives, Mali, Malta, Marshall Islands, Mauritania, Mexico, Monaco, Mongolia, Montenegro, Morocco, Namibia, Nepal, Netherlands, New Zealand, Norway, Palau, Panama, Papua New Guinea, Peru, Poland, Portugal, Republic of Korea, Republic of Moldova, Romania, Russian Federation, Rwanda, Samoa, San Marino, Senegal, Serbia, Sierra Leone, Singapore, Slovakia, Slovenia, South Africa, Spain, Swaziland, Sweden, Switzerland, The Former Yugoslav Republic of Macedonia, Tonga, Trinidad and Tobago, Tunisia, United Arab Emirates, United Kingdom of Great Britain and Northern Ireland, United Republic of Tanzania, United States of America, Uruguay, Zambia.

[2] Afghanistan, Angola, Antigua and Barbuda, Barbados, Belize, Brunei Darussalam, Burundi, Cameroon, Cape Verde, Chad, Gambia, Guinea-Bissau, Honduras, Jamaica, Kenya, Liberia, Mauritius, Mozambique, Nigeria, Saint Lucia, Timor-Leste, Togo, Uganda, Ukraine, Viet Nam. As of 23 September 2010, see http://unfccc.int/home/items/5262.php

Table 7.3 *Conditional leadership of industrialized countries (ICs)*

Base year	Country	Reduction commitment (%)	Condition
1990	Belarus	5–10	If Belarus gets assistance from ICs
1990	Croatia	5	To be aligned with EU position on entry into EU
1990	EU-27	**20 unconditional;** 30 in 2020	If other ICs have comparable targets and if DCs contribute adequately
1990	Iceland	30	If other ICs have comparable targets and if DCs contribute adequately
1990	Japan	25	If fair and effective agreement with ambitious targets for all major economies
1990	Liechtenstein	**20 unconditional;** 30 in 2020	If (see EU position)
1990	New Zealand	10–20	If global agreement to keep temperature rise below 2°C; if other ICs make comparable targets and DCs contribute commensurate with their capabilities; effective rules on land use and an effective carbon market
1990	Norway	30 40 in 2020	If global agreement to keep temperature rise below 2°C; and if other major emitting parties reduce emissions
1990	Russia	15–25	If appropriate accounting of Russia's forestry; and commitments by all major emitters
	Switzerland	**20 unconditional;** 30 conditional	If other ICs have comparable targets and if DCs contribute adequately
1990	Ukraine	20	If Annex I countries have targets with 1990 as base year based on Art. 3.13 of KP; Ukraine remains economy in transition country; flexible mechanisms
1992	Kazakhstan	**15 unconditional**	
2000	Australia	**5 unconditional;** 15–25	If ICs and major DCs take action; 450 ppm CO_2-eq or lower global target
2005	Canada	17	Aligned to final US target
2005	USA	Range of 17	If anticipated national energy and climate legislation comes through

Source: Modified and reproduced with permission from Gupta (2011b), *German Yearbook of International Law*, Dunker and Humblot, Berlin.

sought a special status, accepted a target on condition that it would be assisted to achieve it. Kazakhstan, which was not an Annex I country in 1992, now took on a far-reaching unconditional target. The bottom line was that all the developed countries were waiting for the USA (and Australia and Canada) to adopt a binding target, and the USA was waiting for its own Senate to give the green light. Target setting had become captive to USA's domestic politics.

While the USA has been consistently reluctant to adopt targets, Australia and Canada had previously tried to show some kind of leadership (see Table 3.2). One could hope that their present position was only a temporary aberration. The EU continues to show leadership through its unconditional targets – but often these do not include the offshore islands. For example, Denmark's target is exclusive of Greenland and the Faroes.

Developing countries also showed their willingness to adopt targets (see Table 7.4). Singapore and South Korea are very rich countries and there have been expectations that they would take on responsibilities. For other developing countries, this implies that they are willing to move away from the 'leadership paradigm' and take action despite the unwillingness of the long-running largest polluter to demonstrate leadership; that many are adopting policies domestically; and that their conditional approach is a way to leverage action from the USA. They all also refer to Article 4.7 of the Convention, which makes action in developing countries dependent on help from the developed countries.

7.3.2 COP16–18, from Cancun to Doha

COP15 (2009) in Copenhagen and its voluntary and mostly conditional commitments cast a shadow over COP16 (2010) in Cancun. Meanwhile, Wikileaks revealed information that US aid funding to Bolivia and Ecuador was reduced in 2010 allegedly because of their opposition to the Copenhagen Accord. A substantial aid package to the Maldives may have made it more conciliatory (*The US Guardian*, 2010). There were also expectations that the Green Climate Fund would only be primarily loan guarantees and not grants, as most developed countries were facing tight budgets.

COP16 nevertheless adopted the Cancun Agreements emerging from the AWG-LCA. These agreements focus on establishing GHG targets and timetables consistent with the long-term target of keeping emissions below 2°C, in accordance with the principle of common but differentiated responsibilities; and discussing the possibility of a 1.5°C target. They called for deep cuts in emission levels and a global goal for 2050 to be adopted at COP17. They continued to take decisions on other agenda items (see Table 7.5). A part of the Cancun Agreements emerged from the CMP decisions under the AWG-KP. This part recognized that IPCC-III (2007) indicated that Annex I emission reductions could be reduced by 25–40% in 2020 with respect to 1990 levels. It also decided that the base year for the second commitment period would be 1990 but that countries may also use another reference year.

Table 7.4 *Voluntary commitments of selected developing and emerging economies*

Country	Voluntary commitments	In accordance with UNFCCC
Brazil[a]	−36.1% to −38.9% with respect to business-as-usual (BAU) in 2020	Principles, Arts 4(1), 4(7), 10(2)a, 12(1)b, 12(4), CDM
China[b]	−40 to −45% CO_2 emissions per unit GDP by 2020 compared with 2005 level; increase the share of non-fossil fuels in primary energy consumption to around 15% by 2020; increase forest coverage by 40 million hectares and forest stock volume by 1.3 billion m^3 by 2020 from 2005 levels	Principles, Art. 4.7
India[c]	20–25% reduction in emissions intensity of GDP by 2020/2005	Principles, Art. 4(7)
Singapore[d]	16% below BAU in 2020	If global agreement in which all countries implement their commitments in good faith
South Korea[e]	30% below BAU in 2020	
South Africa[f]	34% below BAU in 2020, 42% below BAU in 2025	Principles, Arts 4(1), 4(7), 10(2)a, 12(1)b, 12(4), technology transfer, capacity building, finance
Bhutan	'Bhutan already sequesters more carbon than is emitted.... Our intent is to ensure that our emissions do not exceed our sequestration capacity'	'We will require the support of the international community'
Maldives	'Achieve carbon neutrality as a country by 2020'	The Government 'will register a request for technological, financial and capacity building support for implementation'

[a] http://unfccc.int/files/meetings/application/pdf/brazilcphaccord_app2.pdf
[b] http://unfccc.int/files/meetings/application/pdf/chinacphaccord_app2.pdf
[c] http://unfccc.int/files/meetings/application/pdf/indiacphaccord_app2.pdf
[d] http://unfccc.int/files/meetings/application/pdf/singaporecphaccord.pdf
[e] http://unfccc.int/files/meetings/application/pdf/koreacphaccord_app2.pdf
[f] http://unfccc.int/files/meetings/application/pdf/southafricacphaccord_app2.pdf
Source: Compiled from the formal letters of these countries to the UNFCCC secretariat, and reproduced with permission from Gupta (2011b), *German Yearbook of International Law*, Dunker and Humblot, Berlin.

Cancun was seen as heralding the demise of a legally binding approach in favour of a bottom-up voluntary approach consistent with a more neo-liberal approach. It expanded the options for offsetting, delegated the tasks of raising resources to

Table 7.5 *COP16–17 and CMP6–7 Decisions*

Issue	COP16, Cancun, 2009	CMP6	COP17, Durban, 2010	CMP7
Targets				Socio-eco. impact policies (#5)
Adaptation/impact	Cancun Agreement (#1)	Cancun Agreement (#1)	Guidelines and modalities for NAPAs (#5)	
Land use		LULUCF (#2)		LULUCF (#2)
Methods				Common metrics on CO_2 eq (#4)
FM	AIJ (#8)	CDM guidance (#3), JI (#4), include CCS in CDM (#7)		ET and other FM (#3); CDM (#7) and materiality standard (#9) and CCS (#10), JI (#11)
Finance	GEF review GEF (#2), Guidance to GEF (#3) and LDCD (#5), SCCF Assessment (#4)	AF Board Report (#5), AF Review (#6)	Launched Green Climate Fund (GCF) (#3)	AF Board Report (#6), AF Review (#7)
Other mechanisms		CB (#11)		CB (#15)
NCs	Enhancing New Delhi WP (#7), CB (#10) From ICs (#9)	Supplementary information (#10)		
Compliance Committee (CC)		CC (#13)		CC (#7, #12); Croatia appeal (#14)
Other bodies	Extension of LDC Expert Group (#6)		Established Ad Hoc WG on Durban Platform for Enhanced Action (#1); Outcome of Ad Hoc WG LCA (#2); 4/CP.17 Technology Executive Committee – modalities and procedures	Outcome of Ad Hoc WG on Further Commitments for ICs (#1)
Budget/admin./other	(#11, #12)	(#12), 9/CMP.6 Method for collection of international transaction log fees		(#16, #17)
Amendments		Kazakhstan Proposal on Annex B (#8)		Kazakhstan Proposal on Annex B (#13)

markets, and did not deal with the issue of intellectual property rights in obstructing technology transfer. The REDD decision focused on taking into account social and environmental aspects, as well as including emissions from agriculture and other land-use changes. However, the Cancun decision supports a version of REDD that might marginalize forest dwellers and ignores proposals for a climate justice tribunal.

The Copenhagen and Cancun agreements may lead to a 4–5°C rise in temperature if implemented (Bond, 2011). The G77 negotiator, Lumumba Di-Aping, concluded: 'We have been asked to sign a suicide pact' (cited in Bond, 2011: 7). Some felt that the ideas being marketed would be a 'short-term source of commodification, speculation, and profit' (Bond, 2011: 3).

The Durban COP tried to re-inject enthusiasm into the negotiations. It adopted 19 decisions (see Table 7.5) including one on the Ad Hoc Working Group on the Durban Platform for Enhanced Action. This platform would engage both developed and developing countries in adopting emission targets in a legal instrument to be adopted in 2015 and effective from 2020. It also agreed on a second commitment period for the Kyoto Protocol, so that the existing flexibility mechanisms could continue. Unfortunately, the USA, Russia, Canada and Japan would no longer support a second commitment period, while Australia and New Zealand suggested conditional commitments.

The COP also decided that the Adaptation Committee established at Cancun would oversee the implementation of adaptation under the convention. It launched the Green Climate Fund, which would provide USD 100 billion annually to DCs from 2020 onwards for adaptation and mitigation with a Standing Committee with oversight functions. It also established the Climate Technology Centre and Network.

Since the Bali meeting, there had been discussions on the issue of loss and damage from climate impacts in the AWG-LCA. The developing countries supported institutional arrangements on loss and damage and the AOSIS (2008) proposal for such a mechanism. This included risk assessment, management and prevention; insurance to deal with the risks to agriculture, food security and livelihood; and rehabilitation of displaced people and compensation for their cumulative negative losses. However, most industrialized countries only supported the risk management aspects. COP15 had discussed risk reduction, insurance, loss and damage; COP16 had recommended a work programme to deal with loss and damage linked to climate change; and COP17 now requested the SBI to assess the risk of loss and damage linked with the adverse effects of climate change, approaches to deal with these losses, and the role of the UNFCCC in dealing with these. Meanwhile, the UN International Strategy for Disaster Reduction recommended pre-disaster preparedness such as early warning systems and community evacuation plans; risk

Table 7.6 *COP18 and CMP8 Decisions*

Issue	COP18, Doha, 2012	CMP8
Targets	Advancing the Bali Action Plan (#1) and the Durban Platform (#2)	
Adaptation/impact methods	Loss and damage (#3), NAPA	Methods (#2)
FM	AIJ	Guidance on CDM (#5), JI(#6)
Finance	WP on LT finance (#4), report of Standing Committee (#5), GCF Report (#6), COP-GCF arrangements (#7), GEF Review/Report (#8, #9), LDC Fund(#10)	AF Board Report (#3), AF Review (#4)
Other mechanisms	Doha WP on Art. 6, CB for EITs	CB (#10), CB for EITs (#11)
NCs	Registry, reporting guidelines, NC from Annex I	Supplementary information (#6)
Compliance Committee (CC)		CC report (#12)
Budgetary/ administrative issues/ other		(#13), Int. transaction log fees (#8)
Amendments		New commitment period (#1); proposal by Kazakhstan to include it in Annex B (#9)

reduction measures such as flood protection, building codes, insurance and contingency funds; emergency response measures; and post-disaster rehabilitation to rebuild damaged infrastructure and restore pre-disaster activities.

The CMP discussed many issues (see Table 7.5) but the Durban agreements were seen as 'empty', offering very little new substance (Fuhr *et al.*, 2011). However, it should be noted that by 2011, Parties were reporting nine active GHG emissions trading schemes – the EU, New South Wales, Norway (which began in 2005 and which cooperates with the EU), Alberta (Canada, 2007), New Zealand (2008), Switzerland (2008), North-East USA (2009), UK and Ireland (2010), Tokyo (Japan, 2010). Four more were in development (UNFCCC Compilation, 2011).

In December 2012, COP18 took place in Doha (see Table 7.6). The secretariat opened the meeting arguing that time was running out and 'the door is closing fast on us because the pace and the scale of action is simply not yet where it must be' (Figures in Johnson, 2012). The world was cruising towards a 3–5°C warming if serious action was not taken (UNEP, 2012). Doha launched a second commitment period which began on 1 January 2013 and will continue for 8 years. All the Kyoto mechanisms and accounting rules remain intact for this period but can only be used by Parties to the second commitment period. Parties have agreed to review their commitments by 2014 to see whether it is possible to increase the level of ambition. Their commitments should be in the order of a reduction of 25–40% of 1990 levels by 2020. This outcome has been achieved through an amendment that still has to enter into force, but countries are invited to provisionally apply its articles (see Table 7.7).

A second outcome was the adoption of a timetable for a climate agreement by 2015 that would cover both developed and developing countries from 2020 onwards. Some Parties decided that they would not carry over surplus emission trading credits into the second period (Australia, the EU, Japan, Lichtenstein, Monaco and Switzerland).

A third decision was the adoption of a Climate Technology Centre hosted by UNEP as the mechanism for promoting technology transfer, and a fourth decision was that the Green Climate Fund should begin operations in 2014. In terms of funding, the EU Commission and some EU countries pledged USD 6 billion for up to 2015, and countries were urged to provide funding in the interim period until 2020 and USD 100 billion annually after that. Other decisions included a review of the long-term temperature goal, enhancing adaptive capacity of vulnerable countries, ways to implement NAPAs, a registry of NAMAs and the need to develop new market-based mechanisms. Most developing countries interpret NAMAs as promoting voluntary action which is registered, but whether it is monitored is a sensitive issue. Such action depends partially on assistance received. In total the COP adopted 26 decisions and the CMP 13 decisions (see Table 7.6).

7.4 REDD revisited

Towards the end of Phase 4, the idea of developing an instrument on deforestation within the climate change regime began to take hold. It began as reducing emissions from deforestation (RED) in 2005 and evolved, in Phase 5, to REDD++ (and forest degradation and safeguards to protect local people and ecosystems, Haug and Gupta, 2013). There is still debate about whether such an instrument should be a fund or a market mechanism. Clearly for a market mechanism to work, there is need for binding targets on some countries.

REDD, visualized as a 'win-win' solution, can protect forests, help developed countries achieve their targets efficiently and provide local forest dwellers with resources to compensate them for protecting the forests and their lost opportunity costs if they wished to exploit the forest lands. However, REDD, if not designed well on paper and in the execution phase, may lead to a 'lose-lose' situation (Gupta, 2012a).

Table 7.7 *Proposed amendment to Annex B of the Kyoto Protocol*

Party	QUELRO 2008–2012% of base year/pd	QUELRO 2012–2020% of base year/pd	Ref. year	QUELRO 2012–2020% of ref year	Reduction pledges for 2020% of ref. year
Australia	108	99.5	2000	98	5–15 or 25
Belarus		88	1990	NA	8
Croatia	95	80	NA	NA	20–30
EU*	92	80	1990	NA	20–30
Iceland	110	80	NA	NA	
Kazakhstan		95	1990	95	7
Liechtenstein	92	84	1990	84	20–30
Monaco	92	78	1990	78	30
Norway	101	84	1990	84	40
Switzerland	92	84.2	1990	NA	20–30
Ukraine	100	76	1990	NA	20
Canada	94				
Japan	94				
New Zealand	100				
Russian Fed	100				

QUELRO, Quantified Emission Limitation and Reduction Objective
*The amendment includes a separate entry for each EU country.

Figure 7.1 illustrates the different controversies at different levels. At the lowest agenda level, REDD can (a) prevent deforestation and forest degradation; (b) be a cost-effective solution; (c) generate resources for local people including indigenous people for protecting the forests; and (d) even help local people with their land rights (Griffiths, 2008; Johns *et al.*, 2008). Issues regarding funding, reference levels, monitoring, permanence and additionality can be positive or negative depending on how they are defined. However, by bringing resources this can also (a) heighten the competition between state and non-state actors for control over forest resources to the detriment of the most vulnerable (Griffiths, 2008; Gupta *et al.*, 2013; IUCN, 2010); (b) lead to a relabelling of traditional activities as 'criminal' (McElwee, 2004); (c) lead to elite capture (the capture of resources by the rich) and further marginalization of the weakest, especially in countries with poor governance (Kanowski *et al.*, 2011).

At the medium organizational level, issues of concern include (a) the differential bargaining power of those involved in REDD negotiations along the REDD chain; (b) the fact that while many countries have decentralized forest management issues and transferred power, control and responsibility to the local levels, REDD may confuse if not compromise these processes by bringing in new ideas of vertically integrated nested or embedded forest systems (Phelps *et al.*, 2010); (c) monitoring,

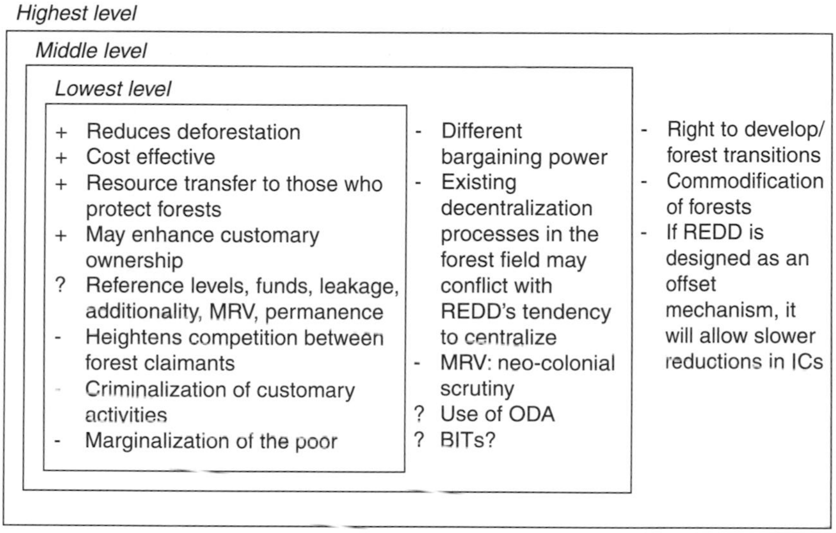

Highest level

Middle level

Lowest level

+ Reduces deforestation	- Different bargaining power	- Right to develop/ forest transitions
+ Cost effective		- Commodification of forests
+ Resource transfer to those who protect forests	- Existing decentralization processes in the forest field may conflict with REDD's tendency to centralize	- If REDD is designed as an offset mechanism, it will allow slower reductions in ICs
+ May enhance customary ownership		
? Reference levels, funds, leakage, additionality, MRV, permanence		
- Heightens competition between forest claimants	- MRV: neo-colonial scrutiny	
- Criminalization of customary activities	? Use of ODA	
- Marginalization of the poor	? BITs?	

Figure 7.1   REDD: Assessing the critique

reporting and verification (MRV) structures needed to ensure the integrity of the financing system may lead to neo-imperialist scrutiny of developing-country activities (Clements, 2010); (d) the possibility that this will absorb ODA resources; and (e) that in the event of the development of a market mechanism, contracts may be drawn up that are subject to bilateral investment treaties.

At the highest ideological discourse level, the worry is that (a) REDD may imply a commodification of forest resources prioritizing carbon content at the cost of other ecosystem services (Clements, 2010; McDermott *et al.*, 2011); (b) and where REDD evolves into a market mechanism, doubts regarding the use of an offset mechanism will once more become critical.

To design REDD to protect forests and local people, the issue of 'leakage' needs to be dealt with. Leakage implies that forest maintenance in one place may lead to deforestation in another. By adopting the state level for accounting purposes, domestic leakage can be avoided (Strassburg *et al.*, 2009). The effectiveness of REDD policies depends on country and forest size as well as de facto control thereover. Effective REDD will require a comprehensive vertically integrated, 'nested' set-up (Cortez *et al.*, 2010; de Gryze and Durschinger, 2010; Pedroni *et al.*, 2009). Identifying the appropriate reference/base level is also critical. Rewarding the prevention of deforestation will initially focus on easy proxy indicators such as the adoption of policies or measures, followed by actual carbon accounting. The problem here is that those who were traditionally deforesting at a high level stand to gain greater resources than those who were protecting their forests. There are possible solutions that can take this problem into account (Parker *et al.*, 2009; Strassburg *et al.*, 2009). Financing REDD can range between 17 and 33 billion

USD annually for halving forest-related emissions (Eliasch, 2008; cf. IWG-IFR, 2009). They could be funded through 'funds', markets or some hybrid combination such as debt-for-nature swaps (i.e. repaying international debts through investing in local nature conservation involving complex financial processes). Fourth, the payments need to be made for verified forest protection and requires reliable monitoring, reporting and verification systems (Grainger and Obersteiner, 2011). Past experience shows that this is not easy (Boyd, 2009); it is easier to evaluate deforestation than forest degradation. Other issues are permanence (Will there be deforestation or forest degradation later in the same area?) and additionality (Is the protection really additional?). In all likelihood, the most critical issue is the issue of safeguards – will forest protection be undertaken in such a manner as to protect the ecosystem services of forests (and not just the carbon content), will it protect or compromise the local communities and indigenous peoples who live in and around the forests (Lyster, 2011) and will it pay symbolic revenues or cover the opportunity costs of avoided deforestation (i.e. deforestation is often undertaken to use the land for profitable uses; not deforesting implies foregoing those profits)? Although an annex was adopted at COP16 in Cancun on safeguards, the question is whether the design of the ultimate regime will account for these or not. The success of REDD depends on the extent to which it can include the multiple equity issues involved, on whether global demand for tropical timber can be tamed, on whether poorer economies can create the sophisticated systems for financial disbursal and monitoring needed to make the system sustainable, and on whether the richer countries/actors can generate the resources needed to make the system viable.

REDD has built up considerable momentum, and relevant policies are being developed in about 40 countries to make them REDD-ready (Angelsen *et al.*, 2009). Many have accordingly developed forest-related policies. In a follow-up to a bilateral agreement with Norway, Indonesia has imposed a two-year moratorium to government-issued licences to convert forest land into other uses. The lack of strong decision making within the climate change regime has not stopped UN-REDD and other actors from pushing the promotion of forest protection further (Agrawal *et al.*, 2011).

7.5 Outside the regime

7.5.1 Recession and climate change

Greenhouse gas emissions are closely linked to economic growth. The economic collapse of the East and Central European countries led to reduced emissions in the late 1980s and throughout the 1990s and 2000s. In 2008, recession hit the developed countries and once more led to a drop in emissions. This could thus be good news in terms of reducing the emissions of developed countries. If targets are not immediately forthcoming, does the recession buy time as global emissions may drop? Research does not provide a clear message. What is clear is that in the early phases of a recession, the economy slows down, unemployment increases and GHG emissions decrease (e.g. Cutlip and Fath, 2012; Zafrilla *et al.*, 2012). Many hope that during the recovery and rebuilding process, governments will promote a 'green economy', decarbonization and sustainability. However, there is a greater likelihood of societies increasing their GHG emissions in the recovery period (Peters *et al.*, 2012) because of: credit shortage (Bloom, 2009); reduced incentive to invest in ET markets as the current EU ETS shows; a shift towards 'growth at all costs' (Senona, 2009); bailouts that don't take environmental issues into account (Reed *et al.*, 2009); a public that prioritizes employment over environmental issues (Scruggs and Benegal, 2012; Shum, 2012); and reduced resources for assistance to developing countries. It might well be wishful thinking that the crisis will lead to scarce resources being invested in green as opposed to business-as-usual options in the absence of far-sighted leadership. Recession has affected the price of credits, but there are already almost ten ET schemes operating worldwide and, as of May 2013, there were almost 7000 registered CDM projects and more than 1,323,545,898 certified emissions reductions issues.

7.5.2 UN-REDD

Although the idea of REDD was launched at COP11, progress was incremental. This spurred initiatives outside the COP and supported ongoing forest governance discussions (Haug and Gupta, 2013). UNEP, the United Nations Development Programme (UNDP) and the Food and Agriculture Organization (FAO) jointly sponsored the UN-REDD initiative in 2007 (UNDP, 2007), building on the strengths of each in forestry. This initiative aimed at helping developing countries prepare to implement REDD – focusing on MRV as well as forest-related capacity building. The World Bank also launched the Forest Carbon Partnership Facility to raise resources and support forest carbon conservation tools and strategies. These two programmes have supported over 50 countries in preparing for carbon sequestration. By early 2012, these had raised about USD 118 million and 232 million, respectively (FCPF, 2011). Additional funds were available for implementing policies in national REDD strategy documents. Furthermore, in 2010, about fifty Parties established the REDD+ partnership for knowledge sharing and learning. These initiatives may fold into the UNFCCC process once it makes significant progress on this issue.

7.5.3 The human rights paradigm: countering market-based approaches

Frustration with the slow progress has led to a revival of the human rights paradigm in relation to climate change. Although mentioned in the Hague Declaration of 1989 (Hague Declaration, 1989), it was only in 2007 that UNDP touched on this issue and a Declaration was adopted in Malé, and in 2008 the UN Human Rights Council decided to analyse the human rights implications of climate change. There have been speeches by UN Human Rights officials also at the COPs since then (see 9.3).

7.5.4 Climate-related agreements in other fora

The climate-related agreements in other fora initiated in the previous period were now sometimes further developed. The International Partnership on the Hydrogen Economy continues to exist and harmonize regulations, codes and standards (RCS) as a way to promote the commercial exploitation of hydrogen and fuel cell technologies. The Global Methane Initiative claims to have reduced 159 million metric tonnes of CO_2 equivalent and has involved some 40 countries and 1100 other partners since 2004. The US government claims that by investing USD 64.5 million, they have leveraged USD 465.8 million from other partners in the process (GMI, 2012). The Carbon Sequestration Leadership Forum has grown since its initial set-up in 2003, and has 24 countries and the European Commission as members. It claims thereby to represent 3.5 billion people and is trying to promote carbon sequestration as a way to address climate change. However, the Asia-Pacific Partnership on Clean Development and Climate completed its activities as of 2011 and no longer exists.

7.6 The problem definition

In Phase 5, the increasing climate scepticism combined with recession has influenced the way climate change is being seen. Global geo-politics appear to be changing with the rise of the BASIC countries – Brazil, South Africa, India and China (and South Korea). This means that there is less money in the developed countries to invest in climate change at home and in terms of assisting developing countries, but it also means that some developing-country emissions are rising rapidly.

In this period, the EU tried to retain its leadership in the climate change process by unilaterally committing to a –20% target and a conditional –30% target. The USA remained a major laggard with a disappointing conditional target. It did not join the second commitment period of the Kyoto Protocol. In an effort to put some constructive vision into the regime, the developing countries proposed their own conditional targets. However, in late 2010, the Canadians decided to follow the US example and the Russians, Japanese and New Zealanders followed suit. Although the secretariat of the Climate Convention under the leadership of Christina Figueres is now pushing three lines of negotiation (under the UNFCCC; under the KP for a second commitment period; and targets for developed and developing countries), it is clear that the developed countries, with the exception of those of the EU, are unlikely to push the regime much further. The question then is whether China and other developing countries will be able to motivate and mobilize the intellectual capacity within their own countries to find the necessary technological and social solutions to addressing the problem. They have put voluntary commitments on the table, although this may not go far enough.

The decline in the power and the willingness of the developed countries is also visible in their efforts to mainstream climate change in development cooperation. This was heavily critiqued by many developing countries who felt that this retracted from the offer of 'new and additional resources' (cf. Government of China, 2008; Government of India, 2008; G77 and China, 2008). Argentina argued thus: 'Argentina sustains that any funding for adaptation must not be counted towards meeting the UN agreed target of 0.7% for aid. Developed countries have delivered just USD 48 million to international funds for least-developed country adaptation, and have counted it as aid. This practice undermines international development and poverty alleviation efforts, upon which current official development assistance is currently founded' (Government of Argentina, 2008). India suggested that the developed countries should provide 0.5% of GNI as new and additional assistance, which was modest in comparison with the G77 and China's request that resources should increase by an additional 0.5–1% of GNI. Much of this perspective was supported by UNDP (2007) and the Commission on Climate Change and Development (2009: 18); although for most developed countries this is not yet convincing.

The offer of USD30 billion in 2012 increasing to 100 billion in 2020 to be raised by new and innovative sources is also viewed with suspicion, as it is not clear who will actually generate these resources and the extent to which they will be loans. The mainstreaming discussion masks two problems – the inability to raise 0.7% for development cooperation and the inability to raise new and additional resources for environmental issues including climate change. Merging the two is instrumental, rational, efficient and possibly effective, but politically very sensitive. The continuous inability to generate the 0.7% amount is seen as reneging on promises made over the last 50 years. The inability to generate the new and additional resources is seen as not taking responsibility for pollution caused primarily in the past by the developed countries and which is adding to the existing burdens of the small countries.

There was also increasing evidence of climate variability (whether climate change or not) and many countries experienced difficulties. Hurricane Sandy in 2012 wreaked devastation in parts of the Eastern USA. There were expectations of increasing migration as a result of climate. The IPCC Special Report on Extreme Events (IPCC, 2012) concluded that disaster losses range from a few billion USD to about 200 billion USD, as was the case with Hurricane Katrina in 2005. Such recorded losses exclude impacts that cannot be easily monetized, such as loss of life or ecosystem services. In the period 2001–2006, middle-income countries lost about 1% of their GDP, low-income countries about 0.3% of their income, and high-income countries about 0.1% of their income. Small island states had lost 1–8% of their income over the period 1970–2010. The report (p. 18) argues that 'The interactions among climate change mitigation, adaptation, and disaster risk management may have a major influence on resilient and sustainable pathways (high agreement, limited evidence). Interactions between the goals of mitigation and adaptation in particular will play out locally, but have global consequences'.

7.7 The role of actors

Although the IPCC (2007) had re-emphasized the seriousness of climate change, the rise of sceptics challenged the integrity of IPCC scientists. The Climate Gate incident of 2009, which resulted from selective exposure of emails from climate scientists from the University of East Anglia, fed public scepticism of climate science and scrutiny of the integrity of scientists in the field.

At the start of this period, all developed countries had put targets on the table. The expectations from the USA under Obama were not met. The withdrawal of other developed countries is disappointing. It remains remarkable that the remaining Annex B countries nevertheless adopted post-Kyoto targets. Although the USA has been recalcitrant in making commitments abroad, some progress domestically was achieved especially at state level.

The G77 (2012: para 59) continued to feel frustrated by the lack of action and reminded the developed countries of their responsibility and of the need to commit to post-2012 commitments. New coalitions were visible in the negotiations. The emergence of BASIC led to reduced G77 unity. The cooperation between AOSIS countries was also under strain. New coalitions between the EU, AOSIS and LDCs became stronger at Durban and Doha. New North–South coalitions were becoming important in this period (e.g. the G20). The Climate and Clean Air Coalition to Reduce Short-Lived Climate Pollutants (CCAC) (e.g. black carbon, methane and some HCFCs) includes 25 developed and developing countries, the European Commission and 23 non-State partners, and was launched in 2012.

Amongst individual actors, indigenous people began to gain in importance. With the UNGA (UNGA, 2007) Declaration on the Rights of Indigenous Peoples and its principle of free, prior and informed consent (FPIC), this group was able to ensure that their needs if not rights were recognized at the COPs (Griffiths, 2008) and a Global Indigenous Peoples Consultation was organized on REDD. They argue against the commoditization of forests that conflicts with their value system.

7.8 Key trends in Phase 5

In this period the EU consistently pushed for targets despite the economic crises in the Eurozone. In doing so the EU has formed a new alliance with the LDCs and AOSIS countries. The domino effect that many had expected in 2001 is gradually beginning to emerge, with Canada, Russia, Japan and New Zealand taking the US position of not accepting targets in the post 2012 period. Fourth, the emerging economies have shown their willingness to adopt commitments despite the fact that basic needs have not been addressed, but there remain enforcement challenges. It is, however, becoming clear that commitments by these countries may not necessarily lead to US commitments. The issue of how this needs to be played out internationally remains a challenge. Finally, the recession – expected to last to 2018 – may lead to short-term decisions (e.g. with respect to extracting shale gas) with long-term lock-in effects.

Part Three

Issues in global climate governance

8

Countries, coalitions, other actors and negotiation challenges

8.1 Introduction

The international climate negotiations reflect complex processes between the various actors. Although it is individual countries that have a vote in the negotiating process, the way these countries form coalitions, the way they are influenced by other actors and the negotiation challenges they face are critical to the way the negotiations develop. This chapter provides a brief flavour of this issue.

8.2 Countries and coalitions

8.2.1 The formal classifications of countries

The climate negotiations involve 194 countries and the EU. Central to the United Nations Framework Convention on Climate Change (UNFCCC) negotiations has been the way countries have been classified (see Chapter 3). The world is divided into developed (Annex I), relatively rich Annex I countries (Annex II) and developing (non-Annex I) countries. This distinction was a cornerstone of the initial agreement based on the idea that the developed countries had polluted more and had greater capabilities for dealing with the problem, and that the developing countries needed room to grow (see 1.4, Table 4.4). The distinction between Annex I and II arose from the argument that the economies in transition (EITs) did not have the resources to transfer to developing countries. The Convention (Art. 4(8)) also developed an elaborate classification of particularly vulnerable countries. In 1997, the Kyoto Protocol created Annex B countries, which included most Annex I countries and others (see Table 8.1).

Annex I was amended at COP3, by consensus, to include Croatia, Liechtenstein, Monaco and Slovenia, and replace Czechoslovakia with the Czech Republic and Slovakia. Any non-Annex I Party can declare to the UN Secretary-General that it intends to be bound by Annex I commitments under the convention. Kazakhstan

Table 8.1 *Evolving classifications of countries*

	Annex I, UNFCCC	Annex B, KP	
	Annex II	EITs	
1992	Australia, Austria, Belgium, Canada, Denmark, EEC, Finland, France, Germany, Greece, Iceland, Ireland, Italy, Japan, Luxembourg, Netherlands, New Zealand, Norway, Portugal, Spain, Sweden, Switzerland, Turkey, UK, USA	Belarus, Bulgaria, Czechoslovakia, Estonia, Hungary, Latvia, Lithuania, Poland, Romania, Russian Federation, Ukraine, UK, USA	
1997	Croatia, The Czech Republic, Slovakia (replacing Czechoslovakia), Liechtenstein, Monaco, Slovenia (by amendment, which came into force in 1998)	Bulgaria, Croatia, Czech Republic, Estonia, Hungary, Latvia, Lithuania, Poland, Rom ania, Russian Federation, Slovakia, Slovenia	Australia, Austria, Belgium, Canada, Denmark, European Community, Finland, France, Germany, Greece, Iceland, Ireland, Italy, Japan, Liechtenstein, Luxembourg, Monaco, Netherlands, New Zealand, Norway, Portugal, Spain, Sweden, Switzerland, Ukraine, UK, USA
2002	Turkey removed from Annex II following amendment (26/CP.7)		
2006			Belarus to be added to Annex B by amendment
2012			USA not included; Belarus and Kazakhstan included. No targets for Canada, Russia, Japan, New Zealand for Kyoto second period

has, since COP3, discussed its possible participation in Annex I and submitted an amendment at COP5. Then Kazakhstan announced that it would consider itself bound by the commitments of Annex I countries. In 2012, Kazakhstan was included in the amendment to the Kyoto Protocol with respect to the second commitment period.

Turkey, an Organisation for Economic Co-operation and Development (OECD) country, was classified as an Annex I/II country in the convention. However, it contested this classification. Azerbaijan and Pakistan proposed an amendment to delete Turkey from both these lists in 1998. It was only in 2002 that Turkey was removed from Annex II following a 2001 amendment. In 2002, Belarus was to be added to Annex B of the Kyoto Protocol.

Two issues arise: first, is it legitimate to differentiate between rich and poor? Clearly, the South of 1990 was likely to suffer significantly from climate change without itself having contributed much to the problem in comparison with the North. The climate problem was defined in the early period to take this into account. However, the legitimacy of this claim has been contested by the US Senate since the early 1990s, although most other countries accepted it.

Second, if legitimate what criteria should one use? When the convention was first designed it did not develop any specific criteria. This led to a rough and ready ad hoc classification based on existing groupings. Singapore was classified as developing while it was very rich, and Latvia and Lithuania were classified as developed while they were in transition. Thus, the group of rich countries (Annex I/Annex B) did not include only rich countries. The group of non-Annex I countries were not just poor countries. And the situation changed considerably over time. Figure 8.1 shows that Qatar, classified as non-Annex I, is now extremely rich. There can be no doubt that Canada, Japan and the USA are very rich countries and have become richer still. However, as no criteria were listed for classification or graduation from one class to another over time, the differentiation has become static.

Criteria for graduation include per capita income and per capita emissions. Such criteria would penalize existing emitters and put others on notice that if their emissions and/or income pass a certain threshold level, they would be expected to take on the responsibilities of countries in that category. However, it should be noted that the economic growth of developed countries (DCs) is not always stable (e.g. the Asian Tigers, Argentina, Iraq) and many developed countries are now in a recession. It should also be noted that the per capita argument is not accepted by those who argue that total, rather than per capita, emissions are more relevant for countries.

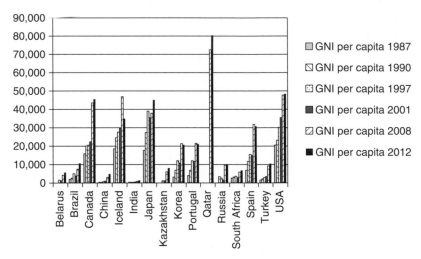

Figure 8.1   Evolving gross national income (GNI) per capita
Source: Generated from the World Bank Data Base.

8.2.2 Negotiation challenges

The climate negotiations have been a learning experience in negotiating theory and practice for most countries. In general, most southern negotiators lack a clear vision about how to develop without polluting; lack relevant knowledge on the evolving agenda items; see climate change as challenging their development; and focus more on the global negotiation challenges than the domestic opportunities and risks. This has, for long, led to a hollow negotiating mandate for the G77 and China – focused more on abstract and general ideas such as the need for capacity building and technology transfer. In the early phases, these countries were unable to rise above their internal differences to generate a common G77 and China position that goes far beyond the lowest common denominator, with political synergy and staying power to put pressure on those developed countries that were unwilling to take action – leading to a handicapped coalition building power. It has been easy for the developed countries, if they so chose, to divide and rule the developing countries. Such divide-and-rule tactics could include bilateral financial incentives and the use of the word 'voluntary' in negotiations. Developing countries vacillate between being pragmatic – 'it's a hard world so take what one can', to being adamant – 'this is too low to accept'. The combination of a hollow mandate and a handicapped coalition power have often led to general, not problem-solving, statements in the international arena (Gupta, 1997; Wagner, 1999). Developing countries have used a defensive negotiating strategy focusing on ad libbing, using proxy indicators of legitimacy such as past ideas and positions from other negotiations, using the power to oppose rather than propose, focusing on damage control, linking climate

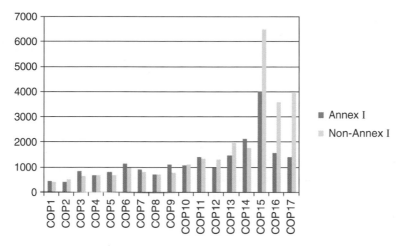

Figure 8.2   Total number of delegates from Annex I and non-Annex I countries
Source: Derived, with permission, from the database of Roger *et al.* (2013).

change to all other international issues, being susceptible to issue-linkages, financial incentives, the threat of the selective use of reprisals in other bilateral relations, and end up feeling cheated by the final results.

The handicapped coalition-forming power makes the defensive strategy brittle, countries choose to form like-minded coalitions with less power and they lack leadership. Besides, the relatively low ability to participate in all the plenaries, formal and informal processes, to cover all issues, and generate the domestic support for national policy has led to a threadbare policy.

Relatively speaking, the developed countries were better prepared. However, even in this group, while the EU has had a well-developed, if inflexible, position, some of the EITs have had significantly weaker positions.

This situation has been changing since 2006. The total number of non-Annex I delegates was larger than that of the developed countries at COP12 and COP13, and from COP15 onwards. This is not remarkable as the developing countries are at least three times more numerous than the developed countries (see Figure 8.2), and thus the average number of delegates is significantly lower. COP15 has seen the highest number of delegates to date (see Figure 8.3). Developing countries are becoming increasingly proactive in the negotiation process and they have had more submissions since 2007 (see Figure 8.4).

8.2.3 The evolution of coalitions

Evolving country coalitions (see Figure 8.5) have also influenced the internal negotiation dynamics. The EU is an institutionalized coalition of increasing numbers of countries (see 8.3.2). The USA mostly speaks for itself (see 8.3.1). Following

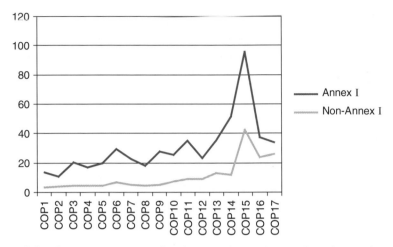

Figure 8.3   Average number of delegates from Annex I and non-Annex I countries
Source: Derived, with permission, from the database of Roger *et al.* (2013).

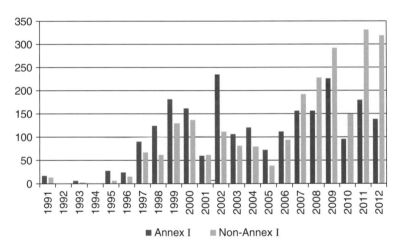

Figure 8.4   Number of submissions from Annex I and non-Annex I countries
Source: Derived, with permission, from the database of Roger *et al.* (2013).

the adoption of the Climate Convention, the USA joined some of the remaining non-EU developed countries to negotiate as the informal JUSSCANNZ (Japan, the USA, Switzerland, Canada, Australia, Norway and New Zealand) group. After 1997, an Umbrella group was established. Some coalitions are formal and inflexible such as the EU, while others are informal and dynamic such as JUSSCANNZ and the Umbrella group. The Central Group 11 was established in 1990. In 2001, the CACAM group was established to include Albania, Armenia, Georgia, Kazakhstan, Turkmenistan, the Republic of Moldova and Uzbekistan. In 2010, a

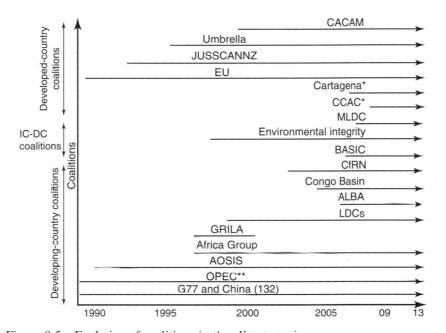

Figure 8.5   Evolution of coalitions in the climate regime
* Not formal negotiating groups; ** OPEC is shorthand for oil exporters in the Convention led by Saudi Arabia; although BASIC has been active since 2009, its first statement was in 2011; GRILA never made plenary statements but was influential as an actor; LDC activity in the climate regime began in 2000; ALBA became active in the climate regime in 2009; the African Group began in the post-Kyoto period but only really became active in 2006. Based on discussions with Nicholas Chan.

Group of Mountain Landlocked Developing Countries was formed by Armenia, Kyrgyzstan and Tajikistan. Their key issues are transportation and food insecurity. In 2012, the Climate and Clean Air Coalition to Reduce Short-Lived Climate Pollutants (CCAC) (i.e. black carbon, methane and some HCFCs) was established.

The developing countries are represented by the G77 and China (see 8.4.1). Interest-based groupings include the Alliance of Small Island States (AOSIS), which focus on their vulnerability to the impacts of climate change (see 8.4.3); the oil exporters in OPEC that focus on their dependence on oil exports and hamper the negotiation of GHG targets (see 8.4.5); the Coalition for Rainforest Nations (Argentina, Bangladesh, Belize, Cameroon, Central African Republic, Chile, Congo, Costa Rica, Côte d'Ivoire, Dominica, Dominican Republic, DR Congo, Ecuador, El Salvador, Equatorial Guinea, Fiji, Gabon, Ghana, Guatemala, Guyana, Honduras, Indonesia, Jamaica, Kenya, Lesotho, Liberia, Madagascar, Malaysia, Nicaragua, Nigeria, Pakistan, Panama, Papua New Guinea, Paraguay, Samoa, Sierra Leone, Solomon Islands, Suriname, Thailand, Uruguay, Uganda, Vanuatu and Vietnam) which focus on protecting rainforests and were instrumental in pushing for reducing emissions from deforestation and forest degradation (REDD). Regional groupings include those involving Africa and Latin America, and in 2004 the Bolivarian Alliance of the Peoples of our Americas (ALBA) was established. The latest addition is BASIC – Brazil, South Africa, India and China.

In 2000, Switzerland, Mexico and South Korea formed one of the first North–South groups – the Environmental Integrity Group – which expanded to include Liechtenstein and Monaco. In 2010, the Cartagena group of 27 countries was established to promote ambitious goals. This group included Antigua and Barbuda, Australia, Bangladesh, Belgium, Colombia, Costa Rica, Ethiopia, France, Germany, Ghana, Indonesia, Malawi, Maldives, Marshall Islands, Mexico, The Netherlands, New Zealand, Norway, Peru, Samoa, Spain, Tanzania, Thailand, Timor-Leste, Uruguay, the UK and the EC.

Figure 8.6 shows the different country coalitions and groupings as of 2012. This reveals the multiple loyalties of the various countries.

8.3 Developed-country actors and coalitions

8.3.1 The USA

Amongst the developed countries, the USA is a key actor. US science has been critical in informing and shaping global understanding of the climate change issue. However, from the start of climate negotiations, the USA has taken a reluctant role in the process. In the run-up to the negotiations, the USA organized a White House Conference in 1990 whose main purpose was possibly to delay action till more science was available. In the period 1990–1992, it was clear that the USA was not willing to accept targets and was opposed to the concept of open-ended principles in the convention. A change of US lifestyle was not acceptable to President Bush. By the time the Clinton–Gore Government came to power, there was more constructive support for the negotiations. The Byrd–Hagel Resolution, however, clarified that the US Senate was unlikely to support far-reaching targets if the developing countries did not also commit themselves to meaningful participation. Nevertheless the US government supported the Kyoto Protocol and helped promote binding quantitative commitments for the developed countries, and pushed for the promotion of market mechanisms. Subsequently, the USA refrained from submitting the protocol to the Senate for ratification and instead attempted at implementing the elements that could be implemented. With George W. Bush coming to power, the USA formally pulled out of the Kyoto Protocol. In the ensuing period, the USA invested in other multilateral agreements with the developing world and others (see 6.4 and 7.4) – seen by some as competing with and distracting from the UNFCCC. As the recession increasingly deepened, US willingness to participate

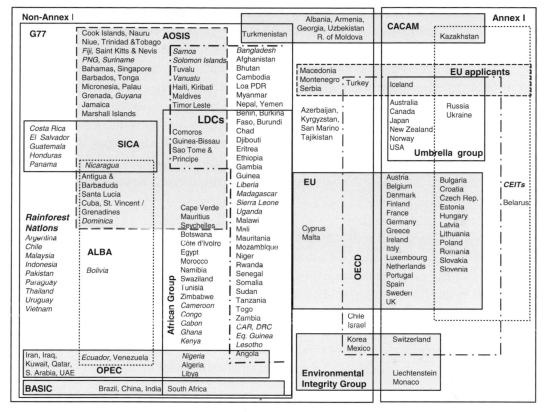

Figure 8.6   Country coalitions and groupings
Source: Adapted and updated with permission from Betzold, C., Paula Castro and Florian Weiler (2012), AOSIS in the UNFCCC Negotiations: From Unity to Fragmentation, *Climate Policy* 12: 591–613.
Note: Croatia has just become a member of the EU in June 2013; Kazakhstan has been included here in Annex I because it has taken on targets for the post-2012 period.

in the negotiations has dwindled further, and they have also opted out of the second Kyoto Protocol period negotiations. However, they are supporting the process towards an agreement which will involve all states (the Durban process). There is no guarantee that the USA will feel obliged to join if such an agreement were to be ultimately adopted. Within the USA there are three dominant frames – scientific scepticism, climate change as a security threat and climate change as an economic opportunity (Fletcher, 2009). It is only the third frame that may help the USA engage more constructively in the climate debate. In the context of the third frame, some action has been taken – for example, at Senate level: the 2003 McCain–Lieberman Climate Stewardship Act, the 2005 Climate Stewardship and Innovation Act, the 2007 Global Warming Pollution Reduction Act and the 2009

American Clean Energy and Security Act (Waxman–Markey). However, none of these have been passed. At provincial and city level, considerable action is being taken (see 8.6.2).

With the discovery of large quantities of shale gas amenable to recovery through hydraulic fracturing (fracking) in the USA, energy security concerns have been allayed while the lock-in into fossil fuels has been further exacerbated. The International Energy Agency argues that the USA may turn into a fossil fuel exporter by 2030, which casts a large shadow over the negotiations. How this plays out in relation to US President Obama's second inaugural address, where he proposed to make the USA a leader in renewable energy, remains to be seen.

8.3.2 The European Union

Contrast the USA's position with that of the EU. The European Community of 12 states steadily evolved into the EU of 15 countries in 1995, 25 in 2004, 27 in 2007, and 28 in 2013. It has tried to demonstrate a leadership role since the beginning. This leadership role in terms of targets is evident in its initial acceptance of a stabilization target for GHG emissions in 2000/1990, its promotion of a −15% target for 2010 and acceptance of a −8% target for the period 2008–2012, and its subsequent promotion of a conditional −30% target and an unconditional −20% target for 2020 (see Table 8.2).

It has been easier for the EU to establish joint common targets than to establish individual targets. It took more than six years to allocate targets among individual member states (Wettestad, 2000). Early efforts to adopt a common carbon tax at EU level failed as countries were afraid to hand over authority over fiscal issues to the Commission (Dahl, 2000). Once the USA left the Kyoto Protocol, the EU renewed its efforts to convince Japan and Russia to ratify it so that it could enter into force. Since then, the EU has intensified efforts to make internal policies on various aspects of climate change. The EU currently has a 20–20–20 strategy focusing on a 20% reduction of GHG emissions in 2020/1990, making renewable energy 20% of total energy, and a 20% increase in energy efficiency.

The EU has an internal emissions trading scheme (EU ETS) that covers emissions from specific sectors emitting about 45% of total GHGs and includes Iceland, Liechtenstein, Norway and Croatia. For those sectors not covered by the EU ETS, an effort-sharing decision has been adopted for 2020 with respect to 2005 (see Table 8.2). This effort sharing, in combination with the 21% reduction in the emissions covered by the EU ETS, should lead to the 20% reduction by 2020. The EU has a soft leadership strategy focusing on directional leadership, which includes diplomacy, persuasion, argumentation and leadership by example. It has been fairly idealistic in pursuing an equitable North–South policy despite hostility from the

Table 8.2 *Evolution of targets of countries that eventually became EU members*

Country	Domestic goal in the early 1990s[a] 1992 Willingness to accept a common stabilization target for 2000/1990	Pre-Kyoto willingness (−15)[b]	1997 Kyoto target Willingness to accept a common −8% target for 2008–2012 for six gases in relation to 1990	Post-Kyoto willingness (−8)[c]	2009 Conditional −30% target for 2030/1990 + unconditional −20% target for 2020/1990	2009[d] effort sharing 2020/2005[4]
Austria	−20% CO_2 in 2000/1988	−25		−13.0	Conditional −30% target for 2030/1990 + unconditional −20% target for 2020/1990	−16
Belgium	−5% CO_2 in 2000/1990	−10		−7.5		−15
Denmark	−20% CO_2 in 2005/1988	−25		−21.0		−20
Finland	Stabilize CO_2 in 2000/1990	−10		0.0		−16
France	Stabilize per capita CO_2 at 2.0 tonnes by 2000	0		0.0		−14
Germany	−25% CO_2 by 25% in 2000/1987	−25		−21.0		−14
Greece	Support EC target	+30		+25.0		−4
Ireland		+15		+13.0		−20
Italy	Stabilise CO_2 in 2000/1990; −20% CO_2 in 2005/1990	−7		−6.5		−13
Luxembourg	Support EC target	−30		−28.0		−20
Netherlands	−3 to −5% CO_2 in 2000/1990	−10		−6.0		−16

Table 8.2 (cont.)

Country	Domestic goal in the early 1990s[a]	Pre-Kyoto willingness (−15)[b]	1997 Kyoto target	Post-Kyoto willingness (−8)[c]	2009	2009[d] effort sharing 2020/2005[4]
	1992					
Portugal	Support EC target	+40		+27.0		+1
Spain		+17		+15.0		−10
Sweden	Support EC/EFTA target	+5		+4.0		−17
UK	Stabilize CO_2 emissions in 2005/1990	−10		−12.5		−16
Bulgaria			−8%			+20
Czech Rep.						+9
Estonia						+11
Hungary						+10
Latvia						+17
Lithuania						+15
Poland						+14
Romania						+19
Slovakia						+13
Slovenia						+4
Cyprus						−5
Malta						+5

[a] Compiled by Wolters et al. (1991).

[b] Council Conclusions of Environment ministers in March 1997 in relation to three gases.

[c] Council Conclusions of Environment ministers in June 1998 in relation to six gases.

[d] Decision No 406/2009/EC of the European Parliament and of the Council of 23 April 2009 on the effort of member states to reduce their greenhouse gas emissions to meet the Community's greenhouse gas emission reduction commitments up to 2020; please note that this effort sharing only covers those sectors not covered by the EU ETS.

USA and Russia with respect to the South. It has a tough balancing act in both promoting trust building with the South, while avoiding the full weight of the 'white man's burden'.

Within the EU there are some interesting story lines developing. The UK Government's 2008 Climate Change Act aims to reduce UK's emissions by 80% in 2050 relative to 1990 levels through four consecutive five-yearly carbon budgets covering the period 2008–2027. These budgets will commit future governments as well, and is a unique step. The German government has had a success story with promoting renewable energy through a strategy to replace coal and nuclear energy and to democratize and decentralize energy. Every German has the right to sell renewable energy to the grid through a feed-in tariff system that guarantees the price for 20 years. Germany now obtains 25–30% of its energy from renewable sources with 50% of the renewable energy owned by its citizens! At the same time, there are chinks in the EU's armour. With the recession and domestic economic challenges, the question is: How long will the EU be able to promote its leadership stance? And given that the EU is increasingly importing goods from other countries, where should the emissions related to these goods be attributed – the producing country or the consuming country?

8.3.3 *The East Bloc*

In the first phase, the eastern and central European countries were communist states. With perestroika in the late 1980s, Russia's control over the Baltic states and the East bloc decreased. Following Tito's death in Yugoslavia and the murder of Ceausescu in Romania, these countries began to undergo political upheaval and economic decline. Their initial aspirations to join the EU led Belarus, Czechoslovakia, Estonia, Hungary, Latvia, Lithuania, Poland, Romania, the Russian Federation and Ukraine to express their willingness to accept a stabilization target in 1992, subject to some leniency with respect to baselines and levels. Later Slovenia and Croatia joined Annex I, but Belarus decided to withdraw, and Czechoslovakia split into two countries. In 1997, it was Bulgaria, the Czech Republic, Estonia, Latvia, Lithuania, Romania, Slovakia and Slovenia that accepted the EU target of −8%, while the remainder negotiated differentiated targets (Poland −6%, Croatia −5%, while Ukraine and Russia negotiated a stabilization target). Some of these countries formed the Central Group 11 (CG11) in 2000 – Bulgaria, Croatia, the Czech Republic, Estonia, Hungary, Latvia, Lithuania, Poland, Romania, Slovakia and Slovenia. This group was later renamed the Central Group, as many of these countries left to join the EU. Although these countries had significant emissions in the past, with their economies collapsing they were able to negotiate their exclusion in Annex II which reduced their responsibilities; they were allowed to use different base years, and later they became eligible for participation in joint implementation, emissions trading and to receive assistance for capacity building.

8.3.4 Russia

The Union of Soviet Socialist Republics was not very active in the pre-1990 phase; thereafter preoccupied with its own political and economic collapse, Russia followed the negotiation process without actively shaping it. As a major producer of GHGs being accountable for about 5% of global emissions, it initially accepted the weakly worded stabilization target in 1992, but by 1997 negotiated a stabilization target for 2008–2012 that allowed it to increase its emissions back to the levels in the 1990 period. As its emissions had fallen by about 34% by 1997, this meant it could increase its emissions by 34% (Nikitina, 2001). The possibility of engaging in emissions trading led Russia (and Ukraine) to join a loose coalition of countries called the Umbrella group. The combination of emissions below the 1990 levels with emissions trading implied that these countries could potentially sell their unused emissions to the developed countries. Given that Russia's emissions in 1990 were estimated at about 17.4% of developed-country emissions, it could make considerable financial gain from participating in emissions trading. Russian industry was eager to participate in the market mechanisms and supported this process. When the USA decided to withdraw from the negotiations, Russia found it had a higher bargaining power as its ratification was essential for the entry into force of the protocol. It was thus able to avail of some issue-linkages including EU support for its entry into the WTO. In 2004 Russia ratified the Kyoto Protocol. Although it was only in 2009 that Russian scientists finally accepted that the climate change problem was serious, the departure of Canada from the Kyoto Protocol inspired Russia too to leave the discussions in 2011. It does have a conditional target on the table of −15 to −25% with respect to 1990 levels, but this is conditional on others taking action and is quite modest. Russia's position has been fairly stable as evident by its cooperation with the Umbrella group as well as through joint statements (Andonova and Alexieva, 2012).

8.3.5 Japan

In the pre-1990 period, Japan was fairly passive on climate change issues. In 1992, it decided to accept the stabilization target for all developed countries. As host of COP3, Japan had the challenging task of ensuring that all developed countries took on targets. During the negotiations in Kyoto, Japan accepted a −6% target, one that would be very difficult to achieve for this already very energy-efficient economy. It was afraid that this would mean further development of nuclear power, an option that was not very popular in Japan. As host to Kyoto, interested in being a global power and wishing to reduce its fossil fuel dependence, Japan led the negotiations in a very positive manner, but was internally very nervous about what the commitment would mean. It adopted Guidelines for Measures to Prevent Global Warming in 1998 and a Law for the Promotion of Measures to prevent Global Warming. It adopted energy conservation measures in industry, transportation, commercial and residential sectors and the promotion of renewable and nuclear energy. It began engaging in voluntary measures with industry. Japan delayed ratification of the Kyoto Protocol. However, once the USA withdrew, the pressure on the remaining countries to ratify increased and Japan ratified the protocol. Since then it had been more or less participating constructively in the process until 2011 when, following Canada and Russia, Japan also announced that it would not participate in the post-Kyoto negotiation process, although it will possibly participate in an agreement that includes all countries.

8.3.6 Others

The remaining countries of Canada, Australia and New Zealand have had varying degrees of commitment in the negotiation period. While Canada and Australia were very ambitious pre-1990 (see Table 3.2), Australia refused to ratify the Kyoto Protocol until a change of government in 2007, and Canada and New Zealand were willing to support the process until their recent change of government, leading them to withdraw from the second commitment period under the Kyoto Protocol.

8.4 Developing-country coalitions and countries

8.4.1 The G77 and China

The developing countries are usually seen as synonymous with the G77 and China. In the climate negotiations, they also include many countries that were initially part of the Second World. Established in 1964 to represent southern interests in international negotiations, the group had grown to 131 countries by 2012. For a long time it had limited resources. However, since 2000, it has had five yearly summits. It promotes 'a solution for the serious global, regional, and local environmental problems facing humanity, based on the recognition of the North's ecological debt and the principle of common but differentiated responsibilities of the developed and developing countries' (G77 and China, 2000). The G77 faces several common sustainability dilemmas which influence the way it negotiates. Its members have similar historical experiences (e.g. colonization), political structures (e.g. weak institutions), economic (have not met basic needs but often have a very rich elite) and technological (e.g. technological followers rather than leaders) conditions. Finally, most have low per capita emissions except Brunei, Qatar and the United Arab Emirates; most have insignificant total GHG emissions except China and India; while for most the impacts of climate change are likely to be relatively high, in physical if not in economic terms. The G77 and China has an annually rotating leadership. Since 1990, the annual chairs have been Bolivia, Ghana, Pakistan, Colombia, Algeria, Philippines, Costa Rica, Tanzania, Indonesia, Guyana, Nigeria, Iran, Venezuela, Morocco, Qatar, Jamaica, South Africa, Pakistan, Antigua and Barbuda, Sudan, Yemen, Argentina, Algeria and, in 2013, it was Fiji. In the last 20 years, OPEC countries have led the G77 six times but AOSIS countries only twice. The first South Summit did not discuss climate change, only biodiversity and desertification. It wanted a just and democratic international economic system and supported the southern development agenda, argued against the use of standards 'as conditionalities for restricting market access or aid and technology flows to developing countries' and claimed that 'we believe that the prevailing modes of production and consumption in the industrialised world are unsustainable and should be changed for they threaten the very survival of the planet'. Finally, 'we advocate a solution for the serious global, regional, and local environmental problems facing humanity, based on the recognition of the North's ecological debt and the principle of common but differentiated responsibilities of the developed and developing countries' (Group of 77 South Summit, 2000). The third Summit in 2012 reiterated the need for quick action.

8.4.2 Africa

Africa has organized itself to participate in UN negotiations since the 1960s, when the African Group of the Whole and the Organization of African Unity were established (Roger and Belliethathan, unpublished). In the pre-1990 phase, African countries were just becoming aware at national government level of the possible impacts of climate change on their countries. In the second phase, 53 African countries mostly negotiated within the G77. But their members were also individually members of AOSIS, least developed countries (LDCs) and OPEC. They formed the African Group of Negotiators (AGN), organized similarly to the African Group of the Whole and including five sub-groups (Northern, Southern, Eastern, Western and Central), and this coalition has negotiated on regional interests since the 1990s. The African countries adopted a common position in 1991 – the African Common Position on the African Environment and Development. The position prioritized poverty reduction, food and energy security and sovereignty in the use of natural resources while requesting the developed countries to implement the precautionary principle in climate change (especially in relation to its impacts on drought and desertification) and to help them with technology and financial transfers. In this phase Africa emphasized desertification, while other developing countries and the developed countries did not do so – which ultimately led to the removal of desertification discussions from the climate negotiations and Africa was promised a separate desertification convention (which emerged in 1994). In the early phases this group had a combination of a relatively low knowledge base, low understanding of their own national interests, limited access to communication resources to enable intra-sessional preparations, an inability to rise above their own differences to develop clear policies, small negotiation teams, unclear mandates, and limited negotiation, agenda setting and coalition-building skills (Grey and Gupta, 2003; Maya and Churie, 1997; Muñoz *et al.*, 2009; Roger and Belliethathan, unpublished). With lower emissions and GDP, Africa's power was also quite limited. The impact of Africa in including desertification and the design of adaptation strategies was quite limited. This may have also been because the content of its strategy did not differ significantly from that of other developing countries and so there was no independent impact. However, in Phase 3, the African group soon realized that equity considerations were not at the heart of the early mechanisms designed, such as the Clean Development Mechanism, and they then developed a common position in 1998 focusing on geographical distribution of projects and equitable geographical representation on the Executive Board. Their distinct views were gradually becoming known although their influence on the negotiations remained minimal (Gupta, 2010a; Roger and Belliethathan, unpublished).

Africa was not benefitting much from adaptation resources nor adequately participating in the CDM; the latter possibly because of the poor market climate, the lack of potential projects, the small size of its energy and industrial sector, the exclusion of deforestation projects in CDM and the exclusion of afforestation projects in the EU's Emission Trading System (Roger and Belliethathan, unpublished). It is only since the fourth phase that the African Group has enhanced its own bargaining strategies and, since the Nairobi COP in 2006, the African Group has been better representing its needs. At Nairobi, the Nairobi Work Programme on Impacts, Vulnerability and Adaptation to Climate Change and the Nairobi Framework was adopted. At Copenhagen the Ethiopian Prime Minister, Meles Zenawi, helped shape the final text on the financial commitment.

This improvement in negotiation style has become possible because of enhanced commitment to negotiate on the greater stakes, increased knowledge based on their own National Communications and NAPAs, enhanced resources for negotiation, the inclusion of non-governmental experts in country delegations, and the establishment of institutional mechanisms (inter-sessional meetings; the development of common positions – e.g. the Algiers Declaration on Climate Change, the establishment of the Conference of African Heads of State and Government on Climate Change (CAHOSCC) in 2009 (which has eight heads of state to help

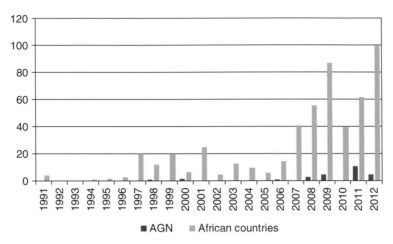

Figure 8.7   Number of submissions and agenda items from African countries and the African Group of Negotiators (AGN)
Source: Derived, with permission, from the database of Roger *et al.* (2013).

develop common African positions) to enhance dialogue between the leaders and negotiators of Africa (Roger and Belliethathan, unpublished). This led to a rise in group and individual submissions of positions on specific issues (see Figure 8.7). Nevertheless, the increasing number of issues and recalcitrance of the developed countries has also led to the disempowerment of this group and its interests in the negotiations (Gupta, 2006; Muñoz *et al.*, 2009).

8.4.3 AOSIS

The Alliance of Small Island States has grown from an informal group of 24 states gathered around the initiative of the Maldives, Vanuatu and Trinidad and Tobago in 1990 to 39 member states, and delegate numbers have risen from 67 in 1995 to 410 at COP16 in 2010. In 1989, the small island states met in the Maldives and adopted the Malé Declaration. They proactively demanded stringent climate measures and assistance for adaptation. The Small Island Developing States (SIDS) is a subset of AOSIS and focuses on the interests of the poorer group within this coalition. The small islands mostly operated individually, if at all, in the first stage of the climate negotiations. But they realized early on that their future was at stake. The group has become influential in terms of making its viewpoint heard, even if it is not always taken into account. This is significant, because apart from its voting power (39 votes), it has a very limited (less than 1%) contribution to global GHGs and GNP (Betzold *et al.*, 2012).

In the second phase, AOSIS lobbied heavily during the climate change negotiations and in 1994 proactively presented a draft protocol (AOSIS Protocol, 1994).

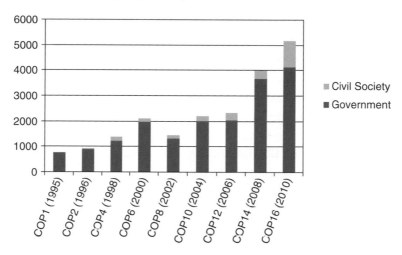

Figure 8.8   Civil society actors versus government actors
Source: Derived, with permission, from the database of Roger *et al.* (2013).

They were not happy with the possibility of instruments that offset emissions in the developed world and lobbied heavily for the Kyoto Protocol. In the third phase, AOSIS made more than 25 written submissions while this decreased somewhat in Phase 4 but increased once more in Phase 5. However, what is noteworthy is that initially member states did not make individual submissions; this changed with a few countries initially making their own submissions (e.g. Dominican Republic, Mauritius) and by Phase 4 significant numbers of countries were making their own submissions (Betzold *et al.*, 2012). The focus of their concern has also changed over time. While initially the focus was on getting mitigation goals for the developed countries and assistance for adaptation, in the third phase they also concentrated on the CDM and land use, land-use change and forestry (LULUCF) issues; the fourth phase on adaptation and a miscellaneous range of issues; and in the fifth phase on mitigation targets for the developed world and REDD. While AOSIS countries are united in the need for a long-term target and for strong mitigation action by the developed countries, their individual interests diverge when it comes to a focus on forests or adaptation issues. For example, in 2009 Tuvalu and Papua New Guinea had diverging views on LULUCF accounting and eligibility under the CDM (Fry, 2008); while Singapore diverged from Tuvalu and Micronesia regarding the relationship between the UNFCCC and the International Civil Aviation Organization and the International Maritime Organization on emissions from aviation and maritime transport, respectively. In terms of forest issues, the conflict is in regard to whether forests and sinks should be used to offset emission reduction in the North or whether such measures help to channel resources to key countries that face deforestation and are better than nothing. This has implied that on the critical issue of REDD, AOSIS was unable to put together a common paper. AOSIS, a tightly organized and cooperative body, is gradually discovering that there are major differences between its member states.

AOSIS has had many successes. Its 1994 Protocol was influential. It created the SIDS – which has a seat in many UNFCCC bodies, including the Bureau, the Executive Board of the Clean Development Mechanism and the Board of the Adaptation and Green Climate Funds, thus breaking away from UN tradition. It has consistently pushed for discussion of a long-term 1.5° target. It is active in the negotiations and uses the negotiation forum to push its views. Its vulnerability to climate change gives legitimacy and strength to its cause, its 'first-mover' advantage, its unified operations, its strong leaders (e.g. Robert van Lierop, Tiuloma Neroni Slade), and its use of international expertise where available has strengthened its global influence (Ashe *et al.*, 1999; Betzold *et al.*, 2012; Davis, 1996; Gupta, 1997; Slade, 2003). However, the only thing that binds AOSIS countries is their vulnerability to sea level rise and extreme weather events. They are very different in culture, spatial distribution, language, forest distribution and their strategies – and even wealth per capita (Kelman and West 2009; Mimura *et al.*, 2007; Wong 2011). This is slowly influencing their ability to shape the debate.

8.4.4 LDCs

Within the G77 is the LDC group, established in 1971 by the United Nations General Assembly (UNGA), which represents the poorest and weakest segment of the international community. It includes 48 countries with about 12% of the world population but having less than 2% of global GDP. Most of these countries are in Africa or are members of AOSIS. Their economic and institutional capabilities are generally low. OPEC's arguments for assistance have stood as an obstruction in the way of assistance to the LDCs, as well as OPEC's resistance on pushing for strong targets, which has affected LDCs' security (Barnett and Dessai, 2002). In the first phase, these countries were unable to significantly push for their interests, but by Phase 3, a special LDC Fund was set up and there was some support for the NAPAs of these countries. However, the LDC fund (LDCF) has had unpredictable and low levels of funding in comparison with what is needed, and is generated primarily from voluntary sources (Davis and Tan, 2010). The funding has been affected by the financial crises, while LDCs themselves have also suffered under the global impacts of these crises. With their weak markets they have also been unable to make adequate use of the CDM (Okubo and Michaelowa, 2010).

8.4.5 OPEC countries

Interest groups within the G77 include the OPEC countries, most of whose economies are dependent on fossil fuel exports. This organization has expanded from 5 member states in 1960 to 12 in 1990. Angola joined in 2007, while Gabon and Indonesia left in 1995 and 2009, respectively, leaving the current membership at 12. OPEC gains from high oil prices while the developing countries lose. However, OPEC does not itself control oil prices which are often more influenced by the four major oil companies that refine and distribute the oil. One should distinguish between OPEC – the organization – and those OPEC countries led by Saudi Arabia that have opposed policies in the developed countries that would lead to a reduced demand for oil, since the developed countries consumed more than 50% of global oil and this would affect their economies. Furthermore, such policies (e.g. carbon taxes) would enable developed countries to gain financially at the cost of the exporters. This led them to include 'the adverse effects of response measures' in the negotiations and demand de facto compensation for lost export revenue. They have continuously blocked progress in the negotiations rather than supported it. Some argue that OPEC and the fossil fuel lobby in the USA have continuously and consistently developed an implicit coalition to protect their own interests.

In 1995, the green G77 was able to isolate the position of the OPEC countries when they insisted on stronger emission reductions in the North. Subsequently, they have ensured that they frequently led the G77 and China and were thus in a position to influence the policies of this group – this despite the fact that G77 and OPEC have little in common. They have been very consistent through the five phases, although in recent years they have also ensured that domestically they can diversify their economies. They have created confusion in the process and diverted attention away from the issue of adaptation and mitigation.

However, OPEC's revenues are not likely to reduce as developing countries have an increasing demand for oil. Furthermore, OPEC will lose if fracking leads to gas exploitation within the developed countries. Diversification is the only way forward for them and many are now investing in that (Barnett, 2008).

8.4.6 China

China is a large country, and is a member of both the Security Council and the G77. In the first phase China argued that the main responsibility for climate change was with the developed countries. It participated actively in the climate negotiations and ratified the convention in January 1993. It was, however, sceptical about joint implementation and in 1996 it submitted a list of technologies that it considered would be appropriate. It participated in the Kyoto Protocol but retained its initial scepticism about the market mechanisms. Later it argued that sinks should be excluded from CDM but that nuclear power should be included. In the meantime the liberalization process set in motion in 1990 gradually had its effect on China. Inefficient small-scale industries in the fossil fuel, aluminium and cement sectors were closed down. China was able to start a decoupling process between emissions and GNP (Zhang, 1999). In the 2000s, the Chinese economy began to take off and with it its GHG emissions, although at a slower pace. China began to actively host CDM projects and hosted the largest number of such projects. China also developed national policies – a national energy efficiency target of decreasing energy intensity (emissions per unit of GDP) by 20% between 2006 and 2010, followed by an additional 16% target for 2011–2015 that should lead to a reduction in carbon intensity of 17%. By the end of 2011, China had the largest amount of installed wind power capacity in the world, and had invested more in hydro and nuclear energy (Xie and Economides, 2009). It has also been able successfully to capitalize on the market mechanism – the CDM. China has now been pushing for targets for the West but has also come up with its own conditional commitment. In the meantme, Chinese emissions have exceeded that of the USA and, as the world's largest producer of emissions, is under considerable pressure also from its own citizens.

8.4.7 India

Another large developing country with a population of more than a billion – India – initially argued that it would first have to raise its standards of living before it could cut down on emissions (Prasad, 1990). It has maintained the argument that its per capita emissions are very low (Dasgupta, 1994). While it initially accepted the Climate Convention, it was sceptical about the use of market mechanisms to offset Western emissions in the Kyoto Protocol (Sharma, 2001), but it ratified the Kyoto Protocol once the USA withdrew. Since then it has begun actively participating in the CDM. The national communications reveal not only that India contributes significantly to GHGs, but will also likely suffer significantly from the impacts of climate change. India argues that it should be allowed to increase its per capita emissions to a level that matches that of the developed countries – as that is the only way it can further develop its economy (Government of India, 2008). However, it has adopted a series of policies on renewable energy and energy efficiency. It intends to reduce its carbon intensity by 20–25% in 2020 with respect to 2005 levels (Walsh *et al.*, 2011). Although officials have been very reluctant to accept a constructive negotiating position, arguing that the USA should take action first, in 2010 India announced its willingness to take on a conditional target. However, for both China and India, it is critical that the common but differentiated responsibilities and respective capabilities principle is articulated and implemented, that there are emissions intensity targets as opposed to fixed levels, that there is funding and technology transfer, and that the non-compliance mechanisms are operationalized with respect to developed countries that have not yet complied with their obligations.

8.5 Other actors

8.5.1 Non-state actors

The participation of non-state actors at the negotiations has grown from 50 at INC1 in Chantilly to 10,000 negotiators and observers in 2000, and the Copenhagen COP of 2009 had 40,000 registered participants; non-governmental organizations (NGOs) have grown from a tiny fraction of national delegations to a substantial number (see Figure 8.8).

Non-state actors include environmental and development NGOs, industry actors, city coalitions, religious groups and scientific institutions, and they are often unable to represent a composite common opinion. The key non-state actor is the Climate Action Network (CAN) with its hundreds of members in its regional networks in all the continents. This organization aims at a common environmental NGO position, which is not easy to attain. Membership includes large environmental NGOs (Greenpeace, World Wide Fund for Nature), prominent southern NGOs like the South Centre in Geneva, the Centre for Science and Environment in New Delhi and the Centre for the Sustainable Development of the Americas.

The epistemic community include the IPCC, top research institutes in the West (e.g. the Foundation for International Environmental Law and Development, the International Institute for Sustainable Development, the World Resources Institute, the Institute for Environmental Studies of Vrije Universiteit Amsterdam and the Stockholm Environment Institute) and from the South such entities as the Bangladesh Centre for Advanced Studies, the Tata Energy Research Institute in New Delhi, the Southern Centre for Energy and Environment in Harare, the ENDA Tiers-Monde in Senegal, etc.

Industry NGOs range from those who favour 'green' action and those who defend existing practices, from the World Business Council for Sustainable Development and its branches, the Business Council for a Sustainable Energy Future, the Insurance Industry Initiative for the Environment, the Pew Center on Global Climate Change (with members like United Technologies, Intel, AEP, DuPont, British Petroleum, Shell, Toyota, Boeing, ABB, Lockheed Martin and Edison International) through the International Climate Change Partnership, the International Chamber of Commerce and the European Roundtable of Industrialists to the Global Climate Coalition, the Coalition for Vehicle Choice and the Climate Council. Although industry was unable to fully pre-empt mandatory targets, they have pushed for carbon markets. This support, and the rise of green business in the mid-1990s, indirectly helped states promote mandatory targets (Meckling, 2011).

Non-governmental organizations have multiple tools at the global level. In addition to lobbying with each other and with negotiators, they try to influence the negotiation process through public awareness and education campaigns both inside and outside of the negotiations. The Earth Negotiations Bulletin (ENB) of the International Institute for Sustainable Development reports factually on all events daily. Critical analysis and perspectives are provided by different journals such as *ECO, Hotspot, CLIME ASIA* and *IMPACT.* The United Nations Non-Governmental Liaison Service produces the Environment and Development File, in which it summarizes key issues in different negotiations. Scientific papers are distributed by the various scientific organizations present. Side-events are organized to bring the latest scientific information to the negotiating table. NGOs may even negotiate on behalf of governments, by joining national negotiating teams. Sometimes in the past they also joined the national delegations of small island states. NGOs are also very active outside the negotiations. They carry out a series of public awareness activities during the negotiations to gain the attention of the public and the media and to collectively put pressure on states to take action. They have activist and advisory roles – some work only within the system and some within and outside (Andresen and Gulbrandsen, 2003).

Much of this activity is unlikely to be very effective if NGOs have not already done their homework in the domestic context of the countries in which they are based. It is at this level that they need not only to raise public awareness, but also to lobby with legislators, undertake advocacy measures with administrators, promote court cases and generate media attention. As time passes, it becomes increasingly clear that there is a huge diversity of perspective between the NGOs (on North–South, green economy versus economic growth, mitigation versus adaptation issues).

8.5.2 *Sub-national authorities*

Sub-national authorities have been engaged in some parts of the world in climate change-related issues for a very long time. In the first phase, some cities (Toronto, Melbourne, some Dutch cities) were organizing meetings to discuss what cities could do. In the second and third phases, there was some degree of slow-down at the city and provincial level as there were expectations that the national governments would show the way. At the same time, activities were institutionalized as part of the International Coalition of Local Environmental Initiatives and the

Climate Alliance of European Cities. However, from the fourth phase onwards there has been enhanced activity. Especially where national governments were particularly recalcitrant, city and provincial authorities began to enter the policy vacuum by pushing for climate change action. This has been the case in the USA and Australia.

In the late 1990s, the state of New York aimed to reduce GHG emissions by 5% in 2010 and 10% in 2020/1990 levels. In 2001, a climate change plan was adopted for Connecticut, Maine, Massachusetts, New Hampshire, Rhode Island, Vermont, Newfoundland, New Brunswick, Nova Scotia, Prince Edward Island and Quebec to reduce their emissions by 10% by 2020/1990 levels. At around that time, about 35 cities in the USA were members of the International Coalition of Local Environmental Initiatives. By 2005, 122 Canadian municipalities, 75 local governments in Europe, 18 in Latin America, 12 in South Africa and 17 in India had joined this coalition. By the mid-2000s, policy processes at local level had intensified, sometimes helped by the national government. As the central government in the Netherlands felt that national policy was less effective, it set up a programme to encourage sub-national authorities to promote climate measures. This programme was able to mobilize more than half the cities in the Netherlands to develop their own climate policies.

There is considerable potential for action at the lowest administrative levels. At these levels, it may be easier to mobilize people and to use spatial planning tools to collectively reduce GHG emissions. These areas can be seen as laboratories of democracy, can fill in a policy vacuum, and can motivate local people to protect themselves (Bulkeley and Betsill, 2003; Bulkeley *et al.*, 2012; Rezessy *et al.*, 2006). As urbanization becomes a dominant element of globalization, policies developed at city level may have an enduring impact on climate change and may help to buy time while national governments try to find a negotiation solution. Having said that, we cannot rely on a bottom-up approach alone as this may not provide the substantial reductions needed; however, these actions may create the necessary public support to ensure national commitment.

8.6 Conclusions: the changing nature of the North–South discourse

An enduring challenge in the climate regime has been the classification without criteria of countries into Annex I and non-Annex I. While the division is both logical and justifiable, the lack of appropriate criteria to guide the graduation of countries to other categories has been an issue of concern. The formal classifications have reinforced the North–South divisions in the climate regime. Movements in the margin – Turkey, Belarus, Kazakhstan – have not established new precedents.

A number of coalitions have also emerged in the process and over time, these reflecting different interests and concerns. The current set of coalitions crosses North–South barriers and has becoming increasingly complex.

Despite the evolving coalitions, although the basic elements of the North–South discourse remain, it is changing over time. The world is evolving from a colonial relationship where the South provided the North with resources, through a phase of dynamic tripartite post-colonial, post-world war ideological politics between the First, Second and Third World, through hardened North–South politics in a world of dominant US unilateralism, to a new geo-political order where there are new powers alongside the old powers creating possibly a new North–South divide between the nouveau riche and the continuing poor. This creates new insecurities in the North about the stability of its own wealth and welfare (which is not set at rest by the current recession) and a greater unwillingness to lead. This nurtures new tensions in the South about the fairness of the international order, as many of these countries still do not quite know how steady their own growth and welfare will be. While the idea of a green economy (i.e. investing in an environmentally friendly economy) beckons, no one knows how to get there.

9

Litigation and human rights

9.1 Introduction

In the pre-negotiation days (Phase 1), early political declarations and articles discussed the issue of 'responsibility' (see Chapter 2). At the time, the concept combined ideas of liability for harm caused to others calling for compensation, with ideas of taking a responsible approach. In Phase 2 of the negotiations and leading up to the adoption of the United Nations Framework Convention on Climate Change (UNFCCC), the use of 'neutral' language became popular. Neutral language is used in the IPCC-III (1990: xxvi) report. Curiously, the report also raised two questions: 'Should mankind's interest in a viable environment be characterized as a fundamental right?' and 'Is there an entitlement not to be subjected, directly or indirectly, to the adverse effects of climate change?' (IPCC-III, 1990: lviii). The text of the Climate Convention also embodied an implicit acceptance that climate change was a global commons problem – which implied a 'neutral factual statement' with no intent to point fingers at individual countries (Bodansky, 1993: 498). The convention adopted the concept of responsibility more in terms of the 'leadership paradigm' (see Chapter 3) and the common but differentiated responsibilities and respective capabilities (CBDR) principle. Responsibility in terms of liability had morphed into responsibility in terms of leadership (Gupta, 1998).

In Phase 3, the perception of the high costs of reducing emissions led to greater emphasis being paid to cost effectiveness as a critical guiding principle and market instruments to promote the implementation of the convention. Leadership was beginning to flag. By Phases 4 and 5, legal options had increasingly become an important issue for discussion and the number of court cases on climate change-related issues began to rise. The issue of human rights also became gradually important (see 9.3). In the future, as the impacts of climate change become increasingly more evident, I think it will be inevitable that legal avenues will be chosen by countries and social actors as a way to promote climate justice. Section 9.2 presents the legal options discussed in the literature and in existing court cases before moving to a more theoretical exploration of the options. The following section discusses human rights issues (see 9.3).

9.2 The role of courts

9.2.1 *Evolution: the rise of literature and legal action*

Since Phase 4 of the climate change negotiation process, there has been an explosive rise in literature on the potential for climate change litigation as one approach to help curb the growth of GHG emissions (Allen, 2003; Anand, 2004; Burns, 2004; Gillespie, 2004; Grossman, 2003; Hancock, 2005; Jacobs, 2005; Lipanovich, 2005; Mank, 2005; Marburg, 2001; Penalver, 1998; Rajamani, 2006; Thackeray, 2004; Verheyen, 2003; Weisslitz, 2002).

The main argument behind this rise in literature is the perspective that the international negotiations are not demonstrating good faith on the part of the (or some) negotiators to address the problem seriously, especially from the perspective of the most vulnerable small island states (Gillespie, 2004). This lack of good faith could lead to a number of legal challenges in different venues. One option is for the small island states to suggest to the UN General Assembly (which requires two-thirds of the members to agree) to request the International Court of Justice (ICJ) for an advisory opinion (legal advice, not a judgement). Particularly vulnerable countries could attempt to sue the USA before the ICJ, but that would require dealing with issues of jurisdiction (the USA may not agree with the ICJ's jurisdiction), wrongful damage to them, and the lack of good faith in implementing the precautionary principle (Jacobs, 2005). The principles of state responsibility and liability for injurious consequences for acts not prohibited by international law could be used to further pursue this issue (Verheyen, 2003). The court may consider the balance of probabilities test to prove causation and may be willing to determine joint and several liability, and future, as opposed to actual, damage. In 2012, Palau announced at the UN General Assembly that it would begin proceedings for requesting the ICJ for an advisory opinion on climate change.

Actions can also be initiated under human rights law (see discussion on human rights, 9.3). Several countries have also requested UNESCO World Heritage status for a number of sites that are particularly at risk from climate change. Nepal, Belize, Peru and Australia, respectively, have requested that Everest National Park, Belize Barrier Reef, Huarascan National Park and the eucalyptus forests of the Blue Mountains in Australia be listed in the List of World Heritage in Danger.[1] This is seen as a way to strengthen the argument that GHG emissions need to be

---

[1] www.climatelaw.org/media/UNESCO.petitions.release.

curbed, otherwise there is a chance that climate change may negatively impact on these protected sites. It may also be a way to seek priority funding for protecting these sites from the impacts of climate change.

Several authors discuss the possibility of considering non-ratification of the Kyoto Protocol by the USA as an 'illegal subsidy' to industry in the USA under the rules of the World Trade Organization (WTO) (Burns, 2004; Doelle, 2004). Non-ratification could also amount to violating the provisions of the International Tribunal for the Law of the Sea for polluting the marine environment under Article 194(2) of the Law of the Sea or under the Straddling Fish Stocks Agreement. At the same time, some see ratification itself as no excuse (see Box 9.1).

Industry could be vulnerable to possible lawsuits in the future. Some argue that it would be wise of industry to voluntarily disclose its GHG emissions to the Securities and Exchange Commission (Hancock, 2005). Some companies have, in fact, started advertising their efforts to reduce GHGs. Others argue that social actors should demand such disclosure from companies. The Australian Climate Action Network has an Australian Climate Justice Program which has given notice, via the law firm Maurice Blackburn Cashman, to the major GHG emitters in Australia about their obligation to address the legal and regulatory risks of climate change (Gupta, 2007). The Carbon Disclosure Project (Carbon Disclosure Project, 2011a; 2011b) has members controlling USD 71 trillion in assets (2011 figure); all members sign up to disclosing their emissions – 91% of Global 500 companies are disclosing their emissions (Carbon Disclosure Project, 2013). They promote the 3% solution, which argues that the US corporate sector could save USD 780 billion by reducing their emissions by 3% between 2010 and 2020.

National courts could also be used by aliens and foreign states. It may be possible for aliens and foreign states to litigate against the US government using US courts (Marburg, 2001). In 2009, Micronesia requested that the Ministry of Environment in the Czech Republic should conduct an EIA of a power plant modernization plan which would lead to GHG emissions in the future.

National courts are more likely to be used by domestic actors. In the USA, there have been court cases requesting the Environment Protection Agency to regulate CO_2 as a pollutant under the Clean Air Act.[2] There have been cases alleging that US government agencies have provided export credits for GHG-producing projects in developing countries.[3] In 2004, litigation was initiated by some US states and

---

[2] 108 Complaint 06–04–03, Commonwealth of Massachusetts, State of Connecticut and State of Maine, plaintiffs versus Christine Todd Whitman, in her capacity as Administrator of the United States Environment Protection Agency.

[3] Friends of the Earth, Greenpeace, Inc. and City of Boulder Colorado versus Overseas Private Investment Corporation, Export-Import Bank of the United States, filed in the US District Court for the Northern District of California, 26 August 2002.

Box 9.1 **Small island states: Attempts at strengthening their legal position**

Small island states made their position clear with respect to the ratification or non-ratification of the Kyoto Protocol by other states in a reservation or comment made at the time of ratification.

For example, the Cook Islands stated that: 'The Government of the Cook Islands declares its understanding that signature and subsequent ratification of the Kyoto Protocol shall in no way constitute a renunciation of any rights under international law concerning State responsibility for the adverse effects of the climate change and that no provision in the Protocol can be interpreted as derogating from principles of general international law. In this regard, the Government of the Cook Islands further declares that, in light of the best available scientific information and assessment on climate change and its impacts, it considers the emissions reduction obligation in article 3 of the Kyoto Protocol to be inadequate to prevent dangerous anthropogenic interference with the climate system.'

The Government of Niue stated that: 'The Government of Niue declares its understanding that ratification of the Kyoto Protocol shall in no way constitute a renunciation of any rights under international law concerning state responsibility for the adverse effects of climate change and that no provisions in the Protocol can be interpreted as derogating from the principles of general international law. In this regard, the Government of Niue further declares that, in light of the best available scientific information and assessment of climate change and impacts, it considers the emissions reduction obligations in Article 3 of the Kyoto Protocol to be inadequate to prevent dangerous anthropogenic interference with the climate system.'

The Island of Kiribati announced that: 'The Government of the Republic of Kiribati declares its understanding that accession to the Kyoto Protocol shall in no way constitute a renunciation of any rights under international law concerning State responsibility for the adverse effects of the climate change and that no provision in the Protocol can be interpreted as derogating from principles of general international law.'

The Republic of Nauru stated that: '… The Government of the Republic of Nauru declares its understanding that the ratification of the Kyoto Protocol shall in no way constitute a renunciation of any rights under international law concerning State responsibility for the adverse effects of climate change; … The Government of the Republic of Nauru further declares that, in the light of the best available scientific information and assessment of climate change and impacts, it considers the emissions of reduction obligations in Article 3 of the Kyoto Protocol to be inadequate to prevent the dangerous anthropogenic interference with the climate system; … [The Government of the Republic of Nauru declares] that no provisions in the Protocol can be interpreted as derogating from the principles of general international law'[.]

Source: http://unfccc.int/kyoto_protocol/status_of_ratification/items/5424txt.php

non-governmental organizations (NGOs) against power companies, arguing that their GHG emissions were a public nuisance.

In Australia, NGOs have tried to prevent mine expansion through law in 2004,[4] and the court ordered in favour of the plaintiffs weighing the present versus future interests of the community. A year later, a legal action was filed against the Australian government for not accounting for the climate impacts on the Great Barrier Reef and the Wet Tropics of Queensland World Heritage Sites. The long delay in ratifying the Kyoto Protocol was seen as violating the World Heritage Convention (Sydney Centre for International and Global Law, (2004).

In Germany, NGOs have argued that export credit provided by the government via Euler Hermes AG helped promote GHG-emitting projects abroad. They requested information disclosure under the Freedom of Information Act. Within the EU there is literature exploring whether states can be held liable when they do not comply with their obligations under the international climate regime and human rights regime (e.g. the European Convention for the Protection of Human Rights and Fundamental Freedoms), and also in general in relation to climate change impacts (Faure and Peeters, eds., 2011). Scholars are examining legal options at supranational level (de Larragán, 2011; Gouritin, 2011; Peeters, 2011), and at national level in the UK (Kaminskaitė-Salters, 2011), USA (Kosolapova, 2011) and the Netherlands (van Dijk, 2011).

In 2003 following the Santa Fe floods, local communities initiated a case against the Government in Argentina for violating Article 6 of the Climate Convention in Argentina. In 2009, the USA followed Canada in including CO_2 as a pollutant. This evolutionary history suggests that there is considerable potential for using litigation as a way to address climate change. But there are a number of hurdles, as the following explains.

9.2.2 The issues

The cases mentioned above can lead to the identification of a number of types of legal action. Figure 9.1 differentiates the possibility of litigation in the international and national arena. It also distinguishes between issues concerning the state and those concerning industry. The types of general legal action are then classified in terms of the different arenas.

One can also argue that at the international level the application of the CBDR principle can give rise to a number of issues (see Figure 9.2). The CBDR principle is quite controversial. This principle recognizes that countries have differential roles and responsibilities with respect to, in this case, the climate problem (French,

---

[4] Australian Conservation Foundation vs. Minister of Planning 2004 VCAT 2029 (29 October 2004).

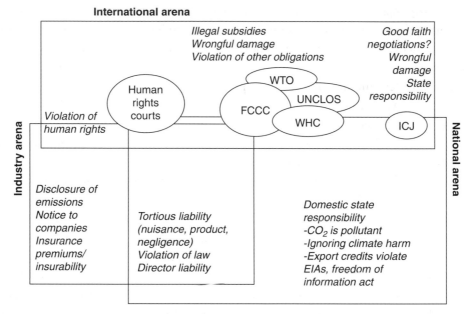

Figure 9.1 (Potential) types of legal action worldwide. WTO, World Trade Organization; FCCC, (United Nations) Framework Convention on Climate Change; UNCLOS, United Nations Convention on the Law of the Sea; WHC, World Heritage Convention; ICJ, International Court of Justice; EIA, environmental impact assessment

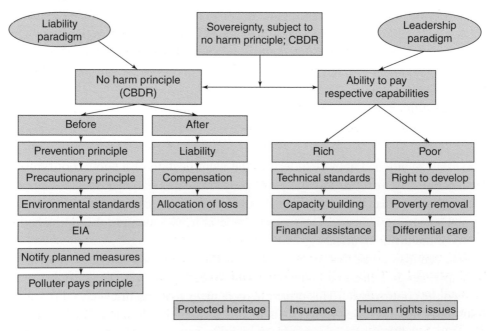

Figure 9.2 Legal options emerging from the CBDR principle. CBDR, common but differentiated responsibilities; EIA, environmental impact assessment
Source: Gupta and Sanchez (2013), p. 28. With permission of Oxford University Press.

2000). Proponents of this principle argue that countries have had varying degrees of contributions to the climate problem and that they have varying abilities to deal with it. This justifies the differentiation (Anand, 2004; Harris, 2001; Honkonen, 2009; Rajamani, 2000). However, others argue that this principle is problematic as it cannot address the magnitude of the problem and that the concept of territorial or per capita variation is not justifiable, because (a) the key problem is total and not per capita emissions and (b) countries are equal sovereign entities (Adams, 2003; Bafundo, 2006). This is especially relevant given the rapid development of the emerging economies. In fact, the size of several countries and their emissions can individually have a substantial effect on addressing the climate problem, and so these countries need to take on additional responsibilities. Finally, since some of these countries have poor governance systems, if action is not taken now they may adopt older technologies that lock their countries into GHG emission-intensive paths (Bafundo, 2006). Although most countries have accepted this principle as it is part of the UNFCCC (UNFCCC, 1992), the fact that the US government is waiting for developing countries to take action before committing to far-reaching action, and that Canada, Russia and Japan have opted out of the post-2012 Kyoto regime, shows that they no longer support this principle.

The above discussion leads to the conclusion that differentiation is only likely to be seen as fair when it is based on clear criteria (Gupta, 2003), when the criteria application is dynamic and allows countries to graduate in and out of responsibilities (cf. Matsui, 2002), and when the criteria are not interpreted as allowing developing countries carte blanche in terms of adopting old and lock-in technologies (cf. Matsui, 2002). However, expecting developing countries to buy new technologies is difficult as these are mostly expensive. This can, to some extent, be addressed through the claim for 'new and additional' resources – a claim that can extend well into the future. However, these resources have tended to be quite limited and focus only on 'incremental costs'. Finally, the application of CBDR should not be seen as implying '*no* responsibility' for developing countries (Mumma and Hodas, 2008: 631). In other words, the developing countries are 'on notice' to take action to avoid GHG-intensive paths where possible. As they become richer, they will have to take on responsibilities.

The CBDR principle is not yet a customary law principle (Stone, 2004). Having discussed the arguments in favour, against and the lessons learned from the CBDR principle, I now argue that this principle can be linked to many existing legal principles (Gupta, 2012a; Gupta and Sanchez, 2012b, 2013). Under international law, each state is sovereign but may not cause significant harm to other states (Rio Declaration, 1992: Principle 2; Stockholm Declaration, 1972: Principle 21; Trail Smelter case, 1941; UNFCCC, 1992: Preamble). Harm is not just how climate change impacts on ecosystem capital and services, sea level rise and extreme weather events, but how this impact affects property (settlements, parks, infrastructure), income (through reduced agricultural and fish yields, loss of working days) and health (thermal stress, spread of disease, impaired nutrition, psychological effects and loss of life) (cf. Smith and Shearman, 2006: 9). In the case of the Climate Convention, this has been translated to some extent in the adoption of the CBDR principle. One can split this principle into two parts – one on responsibilities and one on capabilities. The genesis of the responsibilities principle lies in the 'special responsibility' idea (see Chapter 3) and the liability paradigm. This principle can be divided into implying action before harm takes place and after harm takes place – or after the risk of harm increases. In order to prevent harm, the adoption of the precautionary principle, the principle of environmental standards, environment impact assessments, notifications of planned measures, the polluter pays principle and the user pays principle are critical. Most of these principles are included in the 1992 Rio Declaration on Environment and Development. In terms of 'after harm' or 'before harm is fully manifested', the options are liability for violating the 'before' harm situation, injunctive relief, compensation and the allocation of loss for violating the 'after' harm situation.

In terms of the leadership paradigm (which to some extent covers a non-confrontational responsibility principle), a key element is the ability to pay principle. Here one can differentiate between the rich and the poor. The rich are expected to develop and transfer modern technologies, provide capacity building and help with financial assistance. The poor have differential standards of care with respect to environmental standards, and are allowed to prioritize poverty eradication. These principles emerged from the Rio Declaration and the UNFCCC.

Specific regions of the world are under special protection under the World Heritage Convention and specific communities are under special protection under human rights regimes (see 9.3). Insurance regimes are one way by which countries, groups and individuals may be able to protect themselves from liability and the worst impacts of climate change (see 2.9.2). Figure 9.2 attempts to explain how the CBDR principle can be linked to other principles and concepts.

Although there are many legal avenues at international through to national level, there are a number of legal hurdles (see Figure 9.3). First, who can be a plaintiff? Who does the plaintiff represent? Who can afford to be a plaintiff? Does a plaintiff have standing when a case involves damage to all? It is argued that in line with the US Administrative Procedure Act and the Clean Air Act, any individual who has suffered damages has standing because damage to all cannot be seen as damage to none (Mank, 2005).

Second, a number of actions can be initiated. These can be for preventive or corrective injunctions, demands that the polluter pays, for compensation, even for violating other issues – such as the Freedom of Information Act, the Environment Impact Assessment Law, the breach of a statutory duty or others. The third issue is:

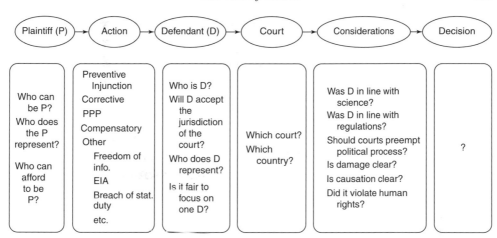

Figure 9.3   The legal process and challenges

Who can be a defendant? Will the defendant accept the jurisdiction of the court? Who does the defendant represent? Is it fair to focus on only one defendant? The fourth issue is: Which court or legal venue is the most relevant, and in which country? The fifth issue is: What are the considerations of the court likely to be? These could include issues such as: Was the defendant acting in accordance with the best available science? Was the defendant acting in line with regulations? Can the court take action when a political process is ongoing? Is there clear damage? Is causation obvious? Does the situation violate human rights? Finally: What is the decision likely to be? (see Figure 9.3).

Much of the critique on the potential of litigation focuses on the issue of causality. However, establishing general causality may not be quite that complicated. A link may be made between Bayesian probability, the IPCC scale, informal scientific scales and legal standards of proof for specific kinds of legal action (Weiss, 2006). For example, a 99% likelihood on the Bayesian scale is more or less equivalent to a 'virtually certain' on the IPCC scale, and could be seen as comparable to 'beyond reasonable doubt', therefore justifying criminal conviction. A little lower down the scale (90–99% probability) is 'virtually certain' on the IPCC scale – and amounts to clear and convincing evidence leading to a quasi-penal civil action. An 80–89% probability is 'likely' and amounts to a 'clear showing', justifying a temporary injunction. Even a 10–22% probability is equivalent to a 'plausible' and 'reasonable indication' and can call for a 'stop and frisk'. However, there is a distinction between general causation and context-specific causality – and this may be much more difficult to prove. Haritz (2010) explains the potential and limits of the precautionary principle as a judicial tool to address uncertainty issues in liability claims. The literature on litigation is increasing rapidly, and the options are also being explored. It may take time, but the overwhelming evidence of knowledge about climate change and its impacts is likely to lead to a situation in which if the carrot of incentives is not sufficient to change human behaviour, the stick of liability may have to be used to move the 'donkey' ahead!

9.3 Human rights and climate change

9.3.1 *The evolution of human rights and climate change*

Initially, there was considerable hope that the problem of climate change would be addressed through the leadership paradigm and that there would be no need to raise the issue of human rights. However, it gradually became clear that the relationship between climate change and human rights needed to be articulated to ensure that there was yet another argument about the need for action on climate change. '[T]he core function of human rights is to counteract gross imbalances of power in society – most commonly those that exist between the state and the individual' (Sinden, 2007: 258; cf. Akin, 2012; cf. Bratspeis, 2011–2012). The recognition of human rights tries to keep the issue of protecting humans central. At the global level there has been a sequential development of rights. Initially, rights focused on political issues (e.g. the International Covenant on Civil and Political Rights; UNGA, 1966a); this requires governments to abstain from certain activities that curtail civil rights. Subsequent rights included social and economic issues (e.g. the International Covenant on Economic, Social and Cultural Rights: UNGA, 1966b); these are rights that require the government to take action to protect individuals. This was followed by a focus on the rights of specific groups – women (The Convention on the Elimination of All Forms of Discrimination Against Women: UNGA, 1979), children (The Convention on the Rights of the Child: UNGA, 1989b) and indigenous people. They also included the emerging rights in the area of specific developmental (Right to Development: UNGA, 1986), sustainable development (UNFCCC, 1992: Art. 3) and environmental issues (Right to Water and Sanitation: UNGA, 2010). In addition, there are other rights including those of refugees. As far back as 1972, the Stockholm Declaration on the Human Environment stated: 'Man has the fundamental right to freedom, equality and adequate conditions of life, in an environment of a quality that permits a life of dignity and well-being' (Stockholm Declaration, 1972: #1). The Declaration of The Hague (Hague Declaration, 1989) stated in its opening paragraph: 'The right to live is the right from which all other rights stem. Guaranteeing this right is the paramount duty of those in charge of all States throughout the world.' It then goes on to argue that climate change threatens the 'very conditions of life'. Justice Weeramantry of the International Court of Justice stated (1997) that protecting the environment is 'a vital part of contemporary human rights doctrine and a *sine qua non* for numerous human rights, such as the right to health and the right to life'. There are also several regional human rights agreements – The American Convention on Human Rights, 1969 and its Protocols; the European Convention on Human Rights, 1950 and its Protocols; the European Social Charter, 1961 and its Protocols; the African Charter on Human and Peoples' Rights, 1981; and the African Charter on the Rights and Welfare of the Child, 1990.

Thus, UN Conventions have, over time, recognized the human rights to life, health, water, food, housing, development and self-determination. They have recognized these rights also with respect to specific vulnerable groups – women, children and indigenous people. Some even argue in favour of a human right to knowledge (Arts, 2009).

Emphasizing the anticipatory and knowledge-driven character of climate change thus shifts the human rights perspective in subtle but significant ways. Rather than focusing on the negative impacts that might occur – important as they are – such a perspective invites us to focus on what we can do to avoid them, through mitigation and adaptation based on equitable universal access to the benefits of science. This is not an alternative to the human rights agenda it is a core component of it.

(Crowley, 2012: 4)

Discussions on human rights and environmental issues received an impetus following the 2002 seminar of the Commissioner for Human Rights and the United Nations Environment Programme (UNEP). The work of the UN Working Group on Indigenous Populations led to a report in 2005. In 2005, local Niger Delta communities filed a suit in the Federal High Court of Nigeria against Shell, Exxon Mobil, Chevron Texaco, TotalFinaELf and Agip, and the Nigerian National Petroleum Corporation[5] requested them to stop flaring gas. This flaring, they argued, violated the fundamental right to life and dignity as recognized by the Nigerian Constitution (Nigerian Constitution, 1999: Arts 33(1) and 34 (1)), the African Charter of Human and Peoples' Rights Act and the Environmental Impact Assessment Act.

In the same year, the Inuit Circumpolar Conference filed a human rights case against the US Bush government for violating their human rights as a consequence of the impacts of global warming on the Arctic. A year later, the commission decided that it was not 'possible to process' the petition as they could not determine on the basis of the facts whether it amounted to a violation of the human rights. The commission did eventually hear the case but did not take action (Knox, 2009).

In 2007, the Maldives (1200 islands, 300,000 inhabitants, average height above sea level 1.5m) hosted a meeting for small islands. The ensuing Malé Declaration

---

[5] Suit No. FHC/CS/B/126/2005; filed in the Federal High Court of Nigeria, in the Benin Judicial Division, Holden at Benin City.

on the Human Dimension of Global Climate Change (2007) stated that: 'climate change has clear and immediate implications for the full enjoyment of human rights'. This declaration requested the Conference of the Parties (COPs) to cooperate with the Human Rights Commission (HRC) and the Office of the High Commissioner on Human Rights (OHCHR) to assess human rights and provide guidance to the COPs.

That year, the UN Human Rights Council adopted a resolution (7/23) stating that: 'climate change poses an immediate and far-reaching threat to people and communities around the world and has implications for the full enjoyment of human rights' as they emerged from human rights documents including the Charter of the United Nations, the Universal Declaration of Human Rights and the International Covenant on Economic, Social and Cultural Rights. This resolution was supported by seventy-eight co-sponsors, including Australia, Germany, India, Indonesia and the UK. Again in 2007, the UN Deputy High Commissioner for Human Rights made speeches at the COPs discussing the link between climate change and human rights. UNDP (UNDP, 2007: 4), in its Human Development Report, stated that climate change is 'a systematic violation of the human rights of the world's poor and future generations and a step back from universal values'.

In 2008, the small island states requested the Human Rights Council to formally request such a report on human rights (Knox, 2009). The HRC adopted such a resolution (2009, 10/4), in which it argued that knowledge about the relationship between human rights and climate change could 'inform and strengthen international and national policymaking in the area of climate change'. At around the same time, the UN Permanent Forum on Indigenous Issues discussed the role of indigenous people, climate change and the environment.

In response to the call for research, the OHCHR (2009, A/HRC/10/61) analysed the documents submitted by several countries and UN agencies and stated that 'climate injustice' was a reality for the most vulnerable countries and peoples of the world – whose contribution to the problem was the least and whose ability to deal with the impacts the lowest. It concluded that climate change impacts would affect access to basic resources – food, water, shelter – needed for survival. However, while it did not 'necessarily violate' human rights (i.e. breach a legal duty), it implied that states had duties to take action on climate change and to cooperate internationally. 'These are not trivial problems, but the OHCHR may overstate the degree to which they prevent a conclusion that at least some effects of climate change violate human rights' (Knox, 2009: 477).

In July 2010, the UN General Assembly adopted the Human Right to Water and Sanitation by 122 votes; this process was fast-forwarded by some developing country governments who felt that the Copenhagen Process was not going fast enough, and they pushed for the recognition of the right and for international responsibilities

(Gupta *et al.*, 2010a). Later that year the Cancun COP also adopted a statement on the link between climate change and human rights, stating that: 'Parties should, in all climate change related actions, fully respect human rights'.

In 2011, the HRC (18/22) announced that there was need for a process to develop and identify how the adverse impacts of climate change impair the enjoyment of human rights, leading to a report for submission to the COPs.

Human rights can be taken into account in the climate negotiation regime through inclusion of human rights provisions, articles on prevention, accountability, reparation and complaints procedures, as well as mechanisms that allow for greater participation of the most vulnerable and the establishment of a subsidiary body that studies these issues and prepares technical papers. The human rights criteria could also be used to assess UNFCCC policies and their implementation (Arts, 2009; Limon, 2009). Human rights can also be taken seriously outside of the climate regime through courts – and indeed European courts have tended to take environmental rights seriously (Kravchenko and Bonine, 2012). Figure 9.4 presents the history of the evolution of the human rights discussion that may be relevant for climate change.

9.3.2 The key issues

There are potentially at least three ways in which human rights can be affected by climate change – on the mitigation side, geo-engineering (kept separate from mitigation because of its different character) and on the adaptation side. The mitigation side refers to the right to develop (UNGA, 1986; see 1.4.2). This right is an 'inalienable human right by virtue of which every human person and all peoples are entitled to participate in, contribute to, and enjoy economic, social, cultural and political development' and includes 'the exercise of their inalienable right to full sovereignty over all their natural wealth and resources'. It includes arguably the possibility to emit GHGs and receive a fair share of the ecospace or climate budget (see Baer *et al.*, 2008; Gupta and van Asselt, 2007). The UNFCCC did not explicitly include this right but hints at it in references to the legitimate needs of developing countries (UNFCCC, 1992: Preambular paras 3 and 23, and Art. 4(5)). It could also refer to the right to sustainable development (see UNFCCC, 1992: Art. 3). '[T]he presumed "right to emit" is clearly something different than the right to vote, the right to receive an education or the right to enjoy basic civil liberties' (UNDP, 2007: 50). However, while a right to pollute is difficult to justify, a right to basic emissions directly or indirectly as a corollary to the right to development is not so easy to dismiss. The HRC Declaration (2011) clearly refers to the right to development (preambular para 2) and reaffirms further the Rio Declaration on Environment and Development, Agenda 21, the Programme for the Further Implementation of

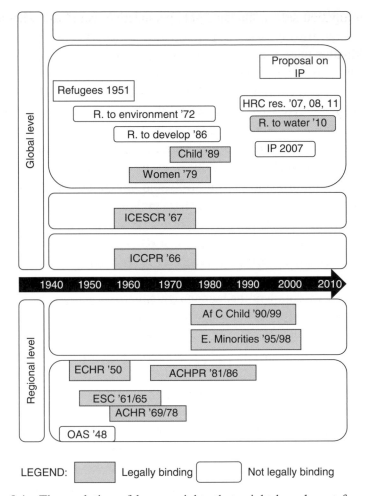

Figure 9.4   The evolution of human rights that might be relevant for climate change.
R, right; IP, indigenous peoples; ACHR, American Convention on Human Rights; ACHPR, African Charter on Human and Peoples' Rights in 1981; AfCC, African Charter on the Rights and Welfare of the Child, 1990; E. Minorities, The Framework Convention for the Protection of National Minorities, Council of Europe; ECHR, European Convention on Human Rights; ESC, European Social Charter; ICCPR, '66, International Covenant on Civil and Political Rights; ICESCR, '67, International Covenant on Economic, Social and Cultural Rights; OAS, '48, American Declaration of the Rights and Duties of Man adopted by the Ninth International Conference of American States, Organization of American States

Agenda 21, the Johannesburg Declaration on Sustainable Development and the Plan of Implementation of the World Summit on Sustainable Development; and recognizes that human beings are at the centre of concerns for sustainable development and that the right to development must be fulfilled so as to equitably meet the developmental and environmental needs of present and future generations.

On the mitigation side, a human rights perspective would ask whether human rights issues require mitigation (and perhaps geo-engineering as a fall-back option), whether mitigation options have human rights impacts and whether market mechanisms have human rights impacts. For example, does the policy process to promote biofuels as a mitigation option lead to a violation of human rights and do mechanisms such as reducing emissions from deforestation and forest degradation (REDD) (see Chapter 7) lead to social conflicts over forest control, also implying a violation of human rights (McInerney-Lankford, 2009)? The adaptation side refers to questions on whether climate change affects the rights of individuals to a home, privacy, property, safe environment, life, food, water and health (Aminzadeh, 2006–2007); whether geo-engineering has an impact on general human rights and, in particular, the rights of specific groups of humans; whether human rights impacts require adaptation policy and policy on receiving refugees; and whether climate change requires an obligation to assist other countries through technology transfer, capacity building and financial assistance (see Figure 9.5).

The development of these rights into legally binding rights requires dealing with some hurdles: first, on the human rights side is the issue of how explicit are these rights and what is the legal status of these rights. There are two schools of thought here – the first argues in favour of limiting human rights to the core principles, while the second argues for expanding the core rights in keeping with the times and new knowledge by either reinterpreting the existing rights in terms of their implications for the environment or creating a specific new category of environmental rights. The latter argument is that some of these rights are very explicitly mentioned in some conventions (i.e. in those on women and the child). There is an explicit UNGA resolution (UNGA, 2010a) about the human right to water and sanitation, which was promoted by countries who were afraid of the possible impacts of climate change on water. The Human Rights Council Resolution (2010) affirmed this right. Second, there is the challenge of establishing causality – the effect of climate change on humans is the result of exposure, sensitivity and adaptive capacity. The inherent vulnerabilities of these actors are included in the adaptive capacity concept. The exposure issue is problematic because of the difficulty of differentiating between climate change and climate variability. Third, there is the challenge of identifying precisely the right holder and the party that is to be held responsible. Would this include all large emitters irrespective of issues of fairness (Knox, 2009)? However, since the causes are 'fragmented temporally and spatially', the remedies are also likely to be fragmented (Akin, 2012: 254). Fourth, is there a significant difference between a rights-based perspective and a social perspective (McInerney-Lankford, 2009)? Fifth, most human rights treaties explicitly make the state responsible for violations within, while climate change calls for both national (as in the Nigeria case) and international claims (cf. UNGA, 2010a; as in the Inuit

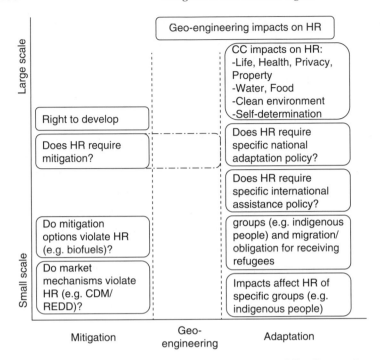

Figure 9.5   Human rights and climate change. CC, climate change; CDM, clean development mechanism; HR, human rights; REDD, reducing emissions from deforestation and forest degradation

people's case). Furthermore, human rights cases normally address past and present violations, while climate change is less about redressing present violations than about preventing future violations. Besides, climate change may also bring benefits and not just violations (US position). Sixth, there is a problem of standing – Who is allowed to bring a claim in a specific court? Seventh, there is the problem of jurisdiction – a so-called responsibility bearer may not accept the jurisdiction of the court (e.g. Inuit case; Aminzadeh, 2006–2007). Other objections include the fact that reinterpreting existing rights will not be easy; that environmental human rights might ignore impacts on ecosystems or detract from other more important human rights issues; and that the rights of future generations are not justiciable. However, Arts (2009) argues that children's rights are well recognized and will most likely be violated by climate impacts, and that there is a clear duty to protect children. Finally, some argue that one should use a positive discount rate in assessing the rights of future generations when comparing these to the rights of current generations, because the current discount rate makes the impacts on future generations far less significant (cf. Caney, 2008, who argues against this viewpoint).

Under human rights law, a rights-based approach to climate change is mandatory (Arts, 2009). It has many advantages: it gives a 'human face' to the problem; focuses attention on generally excluded and marginalized peoples; gives the poor a voice;

levels the playing field between governments and individuals; encourages account-ability, good governance and promotes sustainable outcomes, through human rights standards and principles, including access to information, decision making and a judicial remedy; and provides a greater urgency to the issue. The approach makes sense from a legal and theoretical standpoint because both international human rights law and international environmental law limit the traditional independence and autonomy of state sovereignty (Aminzadeh, 2006–2007: 258; cf. Limon, 2009).

It can thus pierce the recourse to national sovereignty in issues of climate change. It can help in defining the extent to which measures need to be taken on mitigation and adaptation, and can be useful in defining policies on compensation, official development assistance and assistance for migration and refugees. All signatories to the human rights instruments have a responsibility to respect (refrain from interfering with the enjoyment of a right), protect (create rules to prevent violation of these rights) and fulfil (to take measures to ensure these rights) the rights in these instruments.

Is this discussion only creative rhetorical posturing (Sinden, 2007: 258) or more than that? Many argue that despite the legal hurdles, this is a critical option that needs exploring:

And what of the people of the Maldives, the Marshall Islands, Tuvalu, or Vanuatu who risk, because of the economically motivated actions of relatively prosperous people in far-off lands, losing their entire homeland – the country of their birth and the country that their ancestors have inhabited for millennia – and with it their entire culture? Can we tell these people that their human rights have not been violated because it is difficult to apportion responsibility? Perhaps we must, but that is surely because the law is wrong, rather than because our instincts of fairness, equity, and justice are wrong.

(Limon, 2009: 468–9)

The literature argues that sustainable development may call for trade-offs between the economic, social and environmental goals. But precisely in making such trade-offs there is need for guidance on how human rights should be dealt with – such guidance can provide a counterweight to the free-market ideology (Bratspeis, 2011–2012; cf. Akin, 2012). Already the Special Procedures and the Universal Periodic Review of the Human Rights Council are advancing this agenda further (Cameron and Limon, 2012).

9.4 Inferences

The legal literature tends to see climate litigation and legal liability as possible and likely. The rise of legal cases may be just the start of a flood of litigation. The possible threat of such cases may put additional pressure on corporations and governments to be more accountable in the future (Smith and Shearman, 2006): it may serve to prevent future emissions (deterrence), may internalize damage through the polluter pays principle (corrective) and may provide compensation to those affected

(Faure and Peeters, eds., 2011). Legal options and action can take place in the international arena, the industry arena as well as the national arena. They can build on various aspects of the common but differentiated responsibilities principle which themselves can be linked to other principles: the precautionary principle, the polluter pays principle, etc. In the long term some argue in favour of a common liability protocol with clear rules; in fact, industry may even prefer this to changing and different rules in different jurisdictions. However, if there are so many different types of legal actions possible, making a harmonized liability protocol may be quite difficult.

Another related area is the human rights approach. Such an approach is not only beneficial because it focuses on the impacts of climate change on the most marginalized and provides them with a judicial remedy – it may even be mandatory under human rights law. The human rights approach has gone through many generations, and current approaches focus on the rights of special groups and environmental rights. These are being further explored at the global, regional and national levels. This chapter has differentiated between three kinds of ways in which human rights and climate change may interact – in terms of mitigation, adaptation and geo-engineering.

In the meantime, there is a growing social justice movement that is pushing for climate justice worldwide. Many UN bodies – as well as non-UN bodies such as Oxfam International, Mary Robinson's Realizing Rights and Kofi Annan's Global Humanitarian Forum – are now engaging seriously in this discussion. A social movement is being born with the Climate Justice Programme, according to their website: 'a collaboration of lawyers and campaigners around the world encouraging, supporting, and tracking enforcement of the law to combat climate change'. They are tracking the literature and the court cases and/or legal actions worldwide to promote a convergence of legal knowledge and precedent development. Although the 'no-harm' principle has become 'a casualty of the "global commons" framing of the problem' and its phrasing in 'neutral language' (Gupta, 2005), this neutrality is no longer legitimate if there is a lack of good faith in international policy. Haas (1989) argues that epistemic (scientific) communities can shape national and international policies towards a converging goal. Legal scholars interested in social justice issues and litigation options may eventually play a critical role in shaping both national and international science, policy and legal outcomes. Osafsky (2011–2012) argues that environmental justice groups need to frame environmental harm as violating civil rights (and not just as something at the periphery of the social agenda), which will broaden the scope of available remedies and will upscale isolated struggles into a broader agenda.

Part Four

Towards the future

10

Climate governance: a steep learning curve!

10.1 Introduction

Scholars and policymakers have continually argued that the climate change negotiation process is likely to fail. The first sign was evident when US President George Bush was reluctant in the early 1990s, as representative of one of two OECD countries at the time (the other being Turkey), to adopt the GHG emission stabilization target for 2000, stating further that the US lifestyle was not negotiable. This led to the weakening of the climate target in the Climate Convention. Many in the USA saw the convention as a disappointment (Bodanksy, 1993: 454) and as 'environmentally flawed' (Byrd–Hagel Resolution, 1997), because the exemption to developing countries was inconsistent with the need for global action. Subsequent papers predicted that it was most unlikely that a protocol would be adopted at Kyoto; and when the protocol was adopted, that it would not enter into effect. However, the protocol was adopted in Kyoto in 1997 and entered into force in 2005. Although there were high expectations of the Copenhagen COP in 2009, which was to lead to post-Kyoto targets, only a regime of voluntary and conditional commitments emerged. Now that Canada, Russia, Japan and New Zealand have joined the USA in stating that they will not participate in a post-2012 'Kyoto' process, it would appear as if the domino effect has started. The Rio+20 Conference (2012) reminds us that there is a '"significant gap" between pledges and what is needed' and plaintively urges parties to fully implement their commitments. Increasingly there is an expectation that the regime has and will fail (Brand, 2012: 28).

The green and scientific critique is that the convention process is not moving fast enough to address the climate change problem: '[Durban] is an unqualified disaster' and the science and politics are seen as inhabiting 'parallel worlds' (Editorial, 2011). The neo-liberal critique is that the centralized, legalized approach of the Climate Convention does not take into account that it could seriously affect the global economy – that the treaty is 'fatally flawed in fundamental ways' (White

House, 2001). The empiricist critique is that the regime is fragmenting and slow. The climate sceptics argue that the regime is unnecessary.

As submitted in Chapter 1, the key arguments are that climate change is a common-pool problem and there is a powerful incentive for countries to use the resources from the pool without restraint – to free-ride; that people are not willing to change their lifestyles and governments are unwilling to compromise on economic growth; that the global governance system does not allow for, or promote, the cooperation needed to address such a problem; and that there is deep-rooted distrust between countries that lies at the basis of the disagreements between them (Commission on Climate Change and Development, 2009: 9; cf. Opschoor, 2009: 38). Some argue that states are institutionally and culturally locked into the UNFCCC process and that there is 'a stark divide between those who see it as highly successful and others (of the same general ideological persuasion) who see it as a sad and pervasive failure' (Allenby, 2012: 20).

And yet, this chapter argues that the UN process is both necessary and successful. Since Bodansky (2001: 45) reported a decade ago that 'reports of Kyoto's death seem to have been exaggerated', I would submit that the United Nations Framework Convention on Climate Change (UNFCCC) is alive, is learning, evolving and growing in its reach, and that there is a logical and necessary symbiotic process with what goes on outside of the regime.

In comparison with negotiations on fresh water, energy, forests and food, the UNFCCC is the only comprehensive legitimate regime that binds practically all countries in the world (another area possibly being the Law of the Sea). This convention provides the long-term objective; the principles for the allocation of responsibilities; the targets and timetables/pledge and review process; the policies and measures; market mechanisms; the most integrated link with the scientific process; monitoring, review and verification processes; and a compliance system that systematically promotes convention implementation. A judgement of success and failure is always in relation to one's expectations! It is impossible to predict what would have happened in the absence of institutionalized cooperation at global level, there is no counterfactual. However, it is clear that there is far less corrective action in other issue areas that face similar cumulative global consequences.

10.2 Framing

Climate change is a 'glocal problem'. It affects the global climate, which is a heritage of humankind. The climate is something from which no one can be excluded and the use of it by one country or actor does not exclude the use of it by another (i.e. it is a global public goods issue). GHGs mix in the atmosphere – it doesn't matter where the emissions come from, their accumulation leads to enhanced

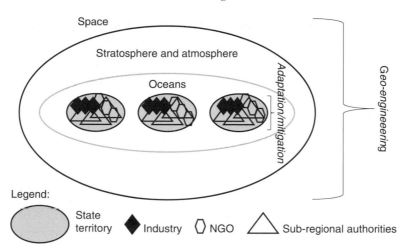

Figure 10.1   The expanding nature of the climate problem. NGO, non-governmental organization

concentrations which can affect the global climate. Climate change is a global 'bad' – GHG emissions lead to climate change which can have all kinds of linear and non-linear impacts on society for which emitters need to be held responsible and 'victims' need to be prepared. Both emissions and impacts will occur at multiple levels of governance and, hence, climate change is a glocal problem. Perhaps climate change is also becoming a 'univerglocal' problem – because some of the measures being discussed involve using space and will undoubtedly have impacts on space (see Figure 10.1)!

If the climate change problem is a common-pool problem, a tragedy of the commons, then the solution, as Garrett Hardin (1968) told us forty years ago, is mutual agreement on mutual coercion. To the extent that climate change is a public bad, markets alone will be unable to address the problem. This means that there is need for multilateral cooperation to create the rules and regulations and a level playing field on which countries can function. The question is: Should the regime focus on depth (i.e. the degree of ambition with regard to global targets) or on breadth (which refers to the extent to which countries may be engaged in the policymaking process)? Some argue for breadth first to ensure legitimacy and learning and negotiating room to allow the small island states and other vulnerable states to push their arguments further, others for depth to ensure efficiency. This trade-off has been the source of enduring debate (Barrett, 2002; Courtois and Haeringer, 2011; Gupta, 1997). Another debate has been on whether negotiations should take place within the context of the UNFCCC or between like-minded countries and in smaller groups. Many see the UNFCCC context as having become fossilized, as repetitive and inflexible and bogged down in North–South politics.

However, I would argue that we are not facing a choice between options, we need multiple mutual agreements at multiple levels of governance and involving multiple sets of actors. Let me briefly explain the role of international treaty making. Some fifty years ago states came together to basically harmonize domestic policies in incremental international agreements. As they were harmonizing existing policies and making incremental improvements at most, there was considerable incentive for states to actually implement the policy. The law-making model has changed considerably since then. Industrialization and globalization have multiplied exponentially the way we change our environment. Now it is science and not existing state practice that provides the guidance regarding the objectives that need to be met and the kinds of measures that need to be taken. This structural model of international law making does not have an inbuilt process of compliance-pull (i.e. have measures that countries have and want to implement). This implies that there needs to be a process by which scientific information and related policy options have to be internalized and socialized by all social actors in society. This calls for a learning process that takes place at multiple levels of governance, especially when the problem being addressed is a glocal, 'wicked' or 'unstructured' challenge.

10.3 Learning

10.3.1 Unstructured problems call for learning

Climate change is an unstructured problem – where scientific consensus is subject to public and political scepticism and there is poor consensus on the values that determine how the problem is to be dealt with (Hisschemöller, 1993). It is a wicked problem with uneven distribution of costs and benefits (Cunningham and Cunningham, 2002). It is a post-normal problem where the science is uncertain, the stakes high and the decisions urgent (Functovicz *et al.*, 1996). Post-normal, wicked and unstructured problems call for a 'learning regime' as a strategy that aims at confrontation, evaluation and integration of diverse and often contradictory viewpoints of a wide diversity of stakeholders at multiple levels of governance (Hisschemöller and Gupta, 1999), which includes participatory integrated assessment (Swart, 1994: 128; Toth, 1995), vision development, lateral thinking, back-casting and transition management (Rotmans *et al.*, 2000), and understanding the politics of governance. It calls for fundamental knowledge co-generation by scientists and society. Contrary to Depledge (2006) who argues that the regime is ossifying, I argue below that there is a learning process and that this is moving quite fast if one compares it to other regimes. The learning literature refers to three orders (first, second and third) of learning.

10.3.2 First-order learning: improving routines

First-order learning is learning how to do something better by, for example, improving one's routines. The climate change treaty process has been going through a complex process of learning. The regular meetings of the Conference of the Parties (COPs) and its subsidiary bodies have led to many incremental changes. Routines have been improved, procedures have evolved and the continuous learning process in the regime has allowed for continuous changes to all of the various aspects of the regime. Although realists would have argued that such improvements are likely to have been only in the interests of the powerful countries, institutionalists argue that problems when divided and sub-divided become manageable because of changing issue-related power configurations providing space for action, because countries become socialized over time, and because ad hoc, no-regrets, cost-effective, incremental and sectoral steps that cost little can be identified and the process may thus inch forward (Benedick, 1993; Greene, 1996; Sand, 1990; Sebenius, 1993). Such disjointed incrementalism has always been seen as an important part of legitimate policy (Braybrooke and Lindblom, 1963).

Initially the process was manageable, because climate change was defined as a single technocratic issue that separated emissions from impacts (and thus reduced the link to liability of the big emitters in relation to those who suffered most from the impacts). Issues such as desertification were moved out of the regime's purview. Other issues such as adaptation were implicitly excluded through a minimal focus. To cope with declining state power, market-based mechanisms were introduced to develop and promote technology transfer to developing countries. An end-of-pipe approach was introduced focusing on emission mitigation and adaptation.

There were and are three major challenges – translation of the long-term objective into a series of short-term consecutive targets; the implementation of the principles; and the step-by-step process to implement the convention. Although the long-term objective was articulated in qualitative terms in the UNFCCC, it took two decades before it was articulated in quantitative terms (see Chapter 6). Despite the initial expectation being that following the initial stabilization target for the developed countries, these countries would systematically reduce their emissions to about 50–60% by 2050, thereby making some room for the emissions of developing countries to grow, this process has clearly suffered setbacks but is not yet lost. There is a range of conditional targets that could easily mature into legally binding targets in the not too distant future. There are five principles adopted in the UNFCCC. I have always argued that two principles were conspicuous by their absence in the Climate Convention – namely the 'no-harm' principle and the 'polluter pays/liability' principles; and cost effectiveness appears to have become more important than the precautionary principle reflecting the growing dominance of the neo-liberal agenda (see Table 10.1).

Table 10.1 *Evolving issues in the climate change regime*

	Pre-1990	1991–1996	1997–2001	2001–2007	2008–2012	Beyond 2012
Problem definition	*Second order learning* Common concern CC as environmental issue; science taken seriously Abstract and future Traditional North–South framing East treated like West	East claims EIT status	CC as development issue; rise of scientific skepticism Present and looming threat		CC in context of recession and recovery, increased scepticism Evolving North–South framing	
Solution framing	Discussion of liability	Leadership ICs reduce emissions and N&A help to DCs	Conditional leadership ICs reduce emissions partly via N&A assistance to South	Leadership competition Mainstream CC in ODA, use markets	Changing leadership in a changing geo-political world Emphasis on human rights	
LT objective	*First order learning* Qualitative target, Quantifying seen as political issue			IPCC provides clarity	Quantitative version adopted	
Targets	Gas-by-gas (CO$_2$)	Stabilization target for 3 gases	Legally binding targets for 6 gases and sinks; offset mechanisms		Voluntary, conditional targets/ NAMAs	
PAMs	Mostly mitigation	Mitigation and AIJ	Mitigation/AIJ/JI/CDM/ET	Forestry/NAMAs/ biofuels	REDD+	
			More attention to adaptation	**Adaptation**	**Adaptation and human rights geoengineering**	**?**
Funds		**IPCC trust fund** **GEF**		**SCCF, LDC Fund, Adaptation Fund, GCF**		

Source: Building on Gupta (2010c). This material is reproduced with permission of John Wiley & Sons, Inc.

The climate regime has inched forward relying on the scientific leadership of the IPCC, the political leadership of the EU, and the moral reminders of the vulnerable countries and peoples. The UK has demonstrated that it is possible to develop emission reduction targets that try to bind future governments and send clear signals to industry (see 8.3.2). The German government has demonstrated that renewable energy can be democratized and decentralized. The Chinese government has demonstrated that it can develop clear signals to sub-national authorities to deal with climate change. It is possible that the global community is on the brink of experiencing the fifth Kondratiev (technological) revolution in combination with a consumption revolution that changes the way society functions.

The climate regime is evolving in the direction of administrative law. This means that the process moves forward even without the explicit consent of national parliaments. In doing so, the process is continuously evolving. The relentless movement of the international process has unleashed national implementation processes across the world and is perhaps to a large extent responsible for national actions in the developing world!

Expanding the negotiating pie is one way forward out of a negotiation deadlock. One route chosen is the UNFCCC route via the Ad Hoc Working Group on Long-term, Cooperative Action under the Climate Convention (AWG-LCA). This route aims to keep the USA on board the negotiation processes; it sees the Copenhagen pledges as part of COP decisions, promoting funding, technology transfer and reducing emissions from deforestation and forest degradation (REDD). A second route allows for further development of the Kyoto Protocol process through promoting a second commitment period (AWG-KP). However, Canada, Russia, Japan and New Zealand have now decided against supporting this route. A third route being investigated is the Durban route Ad Hoc Working Group on the Durban Platform for Enhanced Action (AWG-DP), which aims at a follow-up legal document to be adopted in 2015 with quantitative targets for all major emitters effective from 2020 onwards.

The UNFCCC is very capable of, by itself, engaging in first-order learning. To some extent it can deal with second-order learning, but here broad-based consultation with the outside world becomes imperative. Here there is need for questioning the basic assumptions in the treaty and then going beyond that to search for structural and systemic solutions. This leads to a symbiotic relationship with the outside world in a process that seeks scientific embeddedness, social commitment, legitimacy and effectiveness.

10.3.3 Second-order learning: focusing on proximate drivers and impacts

There is an iterative process between what happens outside the UNFCCC regime and what happens within the regime. The growing evidence outside the regime and the accompanying political commitment led to the climate negotiations in 1990 (see Chapter 3). The difficulties in implementing the convention, the fear that the inverted U-curve did not exist for climate change, and the explicit opposition of the USA to targets juxtaposed against the explicit perception of the EU that targets were necessary led to a situation in which the Kyoto Protocol became possible (see Chapter 5). The continuing fear within developed country societies about free riders, the loss of competitiveness, and leakage made countries afraid to take action, while academics talked about the potential of decarbonization, dematerialization and delinking environmental emissions from growth as the way forward. In the margins there were also discussions about liability and human rights (see Chapters 5, 6, 9). All this amounted to a renewed commitment towards a pledge and review process that emerged from the Copenhagen Conference of the Parties.

This iterative process gave rise to second-order learning (meta-learning, meta-cognition, learning about learning), which refers to learning that challenges the basic assumptions, beliefs and values inherent in the first-order learning process. Clearly, opportunistic incremental and sectoral steps are inadequate because they overlook (a) non-incremental policies; (b) whether the incremental policies have unexpected consequences; and (c) whether policy goals can be achieved by incremental measures (Braybrooke and Lindblom, 1963: 114). This led to the start of a second-order learning process that began in the early 2000s.

Second-order learning engages in broad discussion. Such discussion leads to redefinition of issues over time, and corrections are brought in; climate change was increasingly discussed not just as an environmental issue but also as a development challenge, forest issues were explicitly included through REDD, and adaptation and human rights issues were afforded more importance in the second decade of the negotiations.

10.3.4 Triple-order learning: focusing on underlying drivers and impacts

It is now clear that technological and economic mitigation and adaptation strategies deal only with symptoms. To deal with the systemic and structural causes of the problem, fundamental changes to the system are needed (noted as early as 1992 by Gerlach). This calls for re-learning routines, unlearning, transformative triple-order 'learning' strategies. However, the paradox of paradigm changes, transformative learning strategies and post-normal science is that although these are needed when decisions are urgent, there is no quick way to generate a new global consensus and paradigm changes in a democratic world. This means that even if the locus of formal decision making remains the Climate Convention, there needs to be a pause to generate the global consensus on the nature of the problem and the way in which it needs to be addressed.

Learning in the climate change regime can be illustrated (see Figure 10.2a). In the first stage, there is incremental design improvement to specific instruments (e.g. the CDM, EU-ETS) to enhance their environmental effectiveness. It then becomes necessary to link up with other environmental issues – deforestation, desertification, biodiversity and other land-use issues, water, energy, food, etc. It then becomes apparent that just tinkering with environmental aspects without taking into account development, trade and investment regimes – which, in fact, are a major driver of global change – is inadequate. This calls for second-order learning and a pursuit of coherence in the different arenas. However, merely modifying these other regimes to include sustainability goals is not enough; actual implementation is captive to the way that global society defines progress, structures its production, consumption and distribution patterns, and develops its social and technological infrastructure. This calls for a societal revolution and takes time. At the same time, policies adopted at global and or national level can help to structure the learning process and can be mutually supportive.

Figure 10.2b shows the same information differently. Here the climate change negotiation process leads to advice for policy implementation such as the development of national emission inventories and communications or reports. This two-way process is quite useful initially as it helps to ensure that policies are based on national reports, and national reports cover the key issues mentioned in the policies. But increasingly there is need for engaging other social actors in the policy process. This process of social engagement at global through to local level leads to redefinition of the problem through second- and third-order learning.

This implies that the climate change negotiation process needs a control tower that is moving the regime towards its long-term goals; but it also needs to actively engage social actors in the process of determining how the problem should be addressed.

These two processes of large-scale engagement of social actors, while linking up with other social and economic processes in society, together lead to the search for the paradigm shift. In Chapter 2 it was argued that there are seven levels (see Figure 2.1) at which the problem could be addressed. The highest level was at the level of the paradigm shift, which is essentially towards the notion of a sustainable society. The elusive nature of a sustainable society has led to the concept of sustainable development being increasingly unpackaged and repackaged as green growth, green society and inclusive growth (see Figure 2.2), where each of these paradigms clearly promotes two or more of the three cornerstones of the sustainable development concept (see Section 2.4). The hope is that these paradigms together may stimulate simultaneous demographic, developmental and decoupling (environment from the economy) transitions.

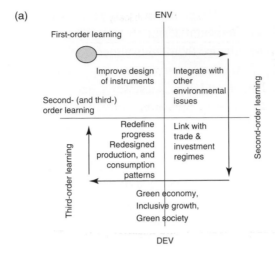

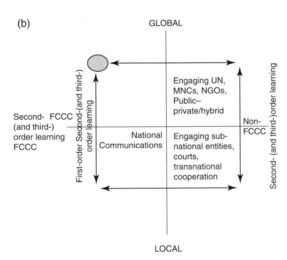

Figure 10.2   First- (a) and second- and third-order (b) learning.
DEV, development; ENV, environment; MNCs, multinational companies; NGOs,
non-governmental organizations; FCCC, (United Nations) Framework Convention
on Climate Change

10.3.5 Implication: coherence where possible, leveraging elsewhere

One could argue that first-order learning translates into improving routines within
the UNFCCC regime; second-order learning translates into revisiting the assumptions underlying the regime and extending outwards towards seeking coherence with
other regimes. This is not always easy. Third-order learning perhaps implies mainstreaming climate change knowledge into our way of living (see Figure 10.3).

In the meantime, it might appear as if climate change is losing ground as a
topic; some call it fragmentation. On the other hand, one could also argue that

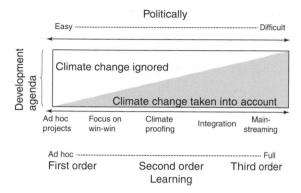

Figure 10.3   First-, second- and third-order learning and their relationship with mainstreaming
Source: Updated and reproduced with permission from Figure 10.2 of Gupta and van der Grijp (2010), Cambridge University Press.

climate change is increasingly being mainstreamed into all UN agencies and UN processes – and in the various ministries in different countries. It calls for transformational learning but also leans on both first- and second-order learning until the transformation is achieved.

Climate science has evolved and new science confirms that climate change is likely to be a serious problem. At the same time the sceptics have organized themselves and, in my view, pollute the public and political debate (see Table 1.1). Although a critical feature distinguishing the last hundred years from previous centuries was the need to make science-based policies, this has given way to stakeholder-based policies, where science is seen as no more than a stakeholder that influences the political process. A growing number of cartoons on the subject seem to represent this shift from respect for, to scepticism of, science.

I would thus make a strong case in favour of promoting the climate change regime as the arena in which science-based targets are developed, national communications required and compliance mechanisms organized. Other issues can be developed and then perhaps exported to other more relevant organizations. For example, if REDD+ is a possible solution to the climate change negotiations, let UN REDD implement this.

10.4 Towards the rule of law at international level for climate governance

The increasing climate scepticism during the current recession, the declining North–South trust and the anticipated impacts of climate change on vulnerable people call for a world order that is seen as fair by a majority of the world's people. This book has discussed the evolving history of the climate change policy process in order to demonstrate that there are very clear reasons why the developing countries feel that in the climate regime, as elsewhere in the UN system, there has been a breach of trust. Let me recall the arguments of the Third World Approaches to International Law school (Chimni, 2006; Khosla, 2007; Okafor, 2005, 2008). This school argues that following the end of colonization and the Second World War, international law was first used by the developed countries to promote their own political post-war interests. Thereafter it was used to develop economic and financial institutions to consolidate their control over the global economic order (Salomon, 2009). Following 9/11, international law has become more and more intrusive (Khosla, 2007). Furthermore, there is a tendency towards institutionalizing a new pragmatism in international environmental law that compromises equity (Gupta and Sanchez, 2012a), adopting a 'West-centric' (Mickelson, 2008) approach that institutionalizes inequities by recommending standard solutions to an unequal world. Universalist arguments need to be unpacked to reveal colonial and neo-colonial ideas (Chimni, 2006; Okafor, 2008; Rajagopal, 2006; Sornarajah, 2001).

The International Council on Human Rights Policy has noted that:

> more than most other issues, climate change throws into relief the inadequacies of the international justice system, given the scale and intimacy of global interdependence that drives the problem and must also drive its solutions.
>
> *(cited in Limon, 2009: 76)*

10.4.1 *The rule of law as applicable to the project of climate governance*

This book has argued that the international climate change regime has gone through five phases and is now at the gates of the sixth phase. During these phases, humanity is learning. In biological terms – this is referred to as learning to survive as a species. The first three phases focused on first-order learning (cognition), improving routines and innovating incrementally. The various innovations in the negotiating process at the UN level demonstrate this. The third to fifth phases have seen the rise of second-order learning (meta-cognition) where some of the fundamental values and assumptions underlying the climate change problem have been questioned and redesigned. It was in this period that climate change was gradually defined more as a development problem rather than simply a technocratic problem, when adaptation was given more emphasis and there was a greater recognition of the need to involve larger numbers of stakeholders in problem solving.

Humanity has now reached a new phase – one in which third-order learning (epistemic learning) is required. Here learning leads society to question its very fundaments. While first- and second-order learning focus on conformative and reformative learning, third-order learning calls for transformative learning (Sterling, 2010–2011). Some of this questioning has led to the development of new paradigms – the green economy, the green society and inclusive growth – all as embodying a further enlargement of the consciousness of the concept of sustainable development. However, this questioning also leads to a re-examination of the fundaments of modern governance systems.

While many are content to accept the limits of international cooperation and rest their hopes on technological optimism and market innovativeness, I would argue for embedding this technological optimism and market innovativeness in a world of good governance and strong international rule of law to support environmental governance, especially with respect to global public goods. This brings me to my final argument. The 20th century saw the end of colonialism and wanton large-scale conquest, and the rise of a legal and economic order to promote peace and development. This led to a counter-movement from the developing world demanding the new international economic order and a movement from many legal scholars – especially from Europe – demanding the rule of law at global level.

The 21st century should bring an end to abject poverty, on the one hand, and limit the role of power play, on the other. It should bring forth a global constitutionalism, good governance and the rule of law at the international level. Good governance is recognition of the shift from government to governance and the need for this new governance pattern to meet the epithet of 'good'. It calls for participatory and democratic governance, transparent, accountable, fair, efficient, effective and responsive governance, governance that observes the rule of law (Botchway, 2001; EC, 2003; Ginther and de Waart 1995; Seif El-Dawla, 2002; UNDP, 1997; Woods, 1999). The rule of law concept generally includes the concepts of generality (applicable to all), publicity (known to all), non-retroactivity (that law should be predictable in advance), clarity (understandable), consistency in design and application (coherence), stability and certainty (not ad hoc), equity and an independent judiciary. A minimal interpretation would promote only procedural rule of law (Craig, 1997: 467; Dworkin, 1985: 165; Esquith, 1999: 335; Fuller, 1964: 106; Hager, 2000: 4; Neumann, 1996: 104; Reynolds, 1989: 3; Tamanaha, 2004). However, this tends to institutionalize the status quo between states. A broader interpretation would promote substantive issues as well – such as equity and democracy (Franck, 1995: 7). The rule of law project levels out the extreme influence of power on politics and is becoming essential for environmental governance.

Born out of the ashes of the world wars, the UN system of organizations and treaties embodied the Kantian dream to create a peaceful world order. Post-war realism and the desire to hold on to some vestiges of power led to a return to a realist framing of international politics (Abbott, 1989; Boyle, 1980) and the rule of law project remained unfinished (UNGA, 2000b). The world has become divided along two groups – those supporting it (Correll, 2001; Foqué, 1998; Henkin, 1969; Kohona, 2002; Kwaka, 1994; Tsagourias, 2003; Watts, 1993) and those opposing it. Arguments against are generally realist arguments that submit that power politics has a critical role in shaping law (Baum, 1986; Georgiev, 1993) and promotes the rules of international law as opposed to the rule of law (Watts, 1993). Arguments in favour of the rule of law are arguments against politics, against the exclusive dominance of power and are, therefore, against the interests of the strong countries and unlikely ever to win ground. Legal theoretical arguments are that the rule of law is based on state practice and thus cannot be an objective standard to evaluate state practice (Koskenniemi, 1990). However, the world is changing so much that it is inevitable that the rule of law project once more comes centre stage for the following reasons.

10.4.2 Global problems multiplying: scale of responsive governance increasing

First, globalization will bring with it more and more 'global commons problems' – problems that cannot be addressed by markets or nation states alone, problems that require nation states to rise above short-term, party political interests to take into account the long-term interests of the global community. Society is already crossing planetary (Rockström *et al.*, 2009) and social boundaries (Raworth, 2012). A European Union Foresight (EC, 2012) project speculates on how the world may develop and what this means for policy. It argues that global population may peak at 9.2 billion around 2075, life expectancy may increase to about 106 from about 2030 onwards, global regional imbalance in population concentrations may increase, urbanization will multiply, more people than jobs will lead to changing employment rules, and economic growth and enhanced global connectivity will lead to greater exhaustion of resources and significantly greater pollution. It concludes that there are different possible evolutions of control over resources – either greater private ownership of these resources regulated by market prices, hegemonic use of power to control extra-territorial resources or a global precautionary principle and a focus on sustainable development. Only the last approach can protect human security globally (see also EC, 2011; MA, 2005: 14–15; Smith, 2010). The scale of the global problem is multiplying; this, in itself, calls for a matching governance response.

10.4.3 Revisiting the fundaments of society: towards predictability

Second- and third-order learning will inevitably make us revisit the fundaments of our production, consumption, distribution patterns; of social ordering. This is already evident in discussions about the green economy, green society and inclusive growth. Revisiting our basic paradigm, unlearning past patterns and searching for new patterns calls for systemic change. Few want to be trend-breakers on their own because of the risks involved. Breaking trends together calls for a predictable set of long-term evolving rules.

10.4.4 Limits of existing governance and incremental innovation

Third, the reactive environmental governance system (Sands, 2003: 620), incoherent policy world (Jackson, 1992), fragmented sectoral arenas and the system of nation states that are not internationally accountable all call for change. The reactive environmental governance system has been continually innovating. Incremental innovation such as the gradual shift to administrative law (Brunneé, 2002; Hey, 2000) can be problematic as it bypasses state consent; the shift to public–private mergers can be problematic as it transforms public into private economic goods subject to confidential contracts and arbitration proceedings (Gupta, 2011a); the shift to legal fragmentation creates inconsistent and unaccountable policy frameworks that together compromise on the fundaments of good governance. International environmental law may be becoming soft – biodiversity, although a hard-law treaty, is increasingly being seen as soft (Harrop and Pritchard, 2011) while the climate regime is moving towards a pledge-and-review regime. If law goes 'soft' it questions the very raison d'être of law. Whose interests are served through such soft targets? The politicization of these regimes leads to a 'pick-and-choose' culture where principles, rules and policies are selected at the will of the most powerful countries at any moment of time. Extrapolating from current governance trends, one can argue that governance will become more tokenistic, opportunistic and simplistic (EC, 2012). While the shift from government to governance in itself is a worthwhile shift, in order to ensure that these shifts are not tokenistic, the rules of governance will have to be created in a way that is both in line with the rule of law and effective (Biermann *et al.*, 2012a, 2012b; cf. Alston, 1997). This may reduce the transaction costs of cooperation.

The incoherence in the policy world reflects the conflicting visions, paradigms, norms and policy instruments chosen by society to deal with specific problems. This can to some extent be dealt with through developing systemic approaches. In water there is a shift from hydraulic and sectoral approaches towards the integrated water resource management paradigm. In ecosystems, there is a shift in thinking towards ecosystem services and in climate change from mitigation and adaptation towards sustainable development. Such systemic thinking, currently also expressed as 'nexus' thinking, is important to help in understanding the links between issues and how best they can be addressed. But systemic thinking does not lend itself to systemic action; action will always focus at specific elements in the system.

Similarly, since environmental problems cannot easily be compartmentalized and are closely associated with trade and investment, if not development, development cooperation, human rights and even military intervention, it is impossible to develop a comprehensive approach for all problems at glocal level. Fragmentation in policymaking is thus inevitable (Gupta and Sanchez, 2012b). However, it is argued by Koskenniemi (2009) that legal fragmentation is not a random result.

Legal fragmentation and the growth of a huge number of subjects (human rights, environmental law, economic law, investment law) that are replacing the common core of international law are the result of competing endeavours by different groups of scholars and actors who wish to promote their own issue-specific interests:

> Through specialization – that is to say, through the creation of special regimes of knowledge and expertise in areas such as 'trade law', 'human rights law', 'environmental law', 'security law', 'international criminal law', 'European law', and so on – the world of legal practice is being sliced up in institutional projects that cater for special audiences with special interests and special ethos.
>
> *(Koskenniemi, 2009: 9)*

He argues further that the search for coherence between these fragmented areas will thus be ephemeral at best, as each arena will promote its own interests and the end outcome will be a lack of coherence reflecting the politics in each arena. This also means that there will be forum shopping – to ensure that issues are decided in specific arenas as opposed to other arenas. If climate change is addressed within the UNFCCC it will be managed by the ethos operating there. If the issue is addressed in other sectors or other fields – a different ethos will emerge. He also states:

> Topics such as 'trade and environment', 'security and human rights', 'development and investment' give name to some such conflicts, while notions such as 'sustainable development', 'responsibility to protect', or 'human security', among a host of others, single out fragile compromises in areas where the struggle between opposing groups of experts and their preferences has not (yet) been taken to the end.
>
> *(Koskenniemi, 2009: 10)*

He opines that political struggles are expressed in the 'neutral language of expertise', which obscures the underlying political choices.

The combination of innovative governance responses involving a multitude of actors, the difficulties in translating systemic thinking into action and the inevitability of fragmentation will all lead to a proliferation of rules (Abbott, 1989; Boyle, 1980). Those with more power and clout may be able to influence the content and style of these rules. However, this proliferation of rules will need a system against which they can be checked for their legitimacy, accountability, equity content, legality, effectiveness and efficiency. It is precisely this fragmentation that will call for the need for a stable rule of law at global level.

Finally, while short-term elected governments are accountable to their electorates/populations, in general they are not accountable internationally or in the long term. This Westphalian system of nation state cooperation has perhaps outlived its usefulness, especially in terms of long-term issues. Sharing one planet inevitably means that global security is intertwined and countries will eventually have to take on obligations *erga omnes* and be willing to be responsible and accountable to each other, as well as domestically.

10.4.5 *Pre-emptive geo-political pragmatism*

The fourth issue brings me to a pragmatic, geo-political argument. A political order based on power protects the powerful. However, there is no guarantee that those in power today will remain in power tomorrow. The current recession in Europe and the USA and the accompanying rise of power of the BASIC (Brazil, South Africa, India and China) economies demonstrates that power is ephemeral at best and can change over time. Global prognosis reveals that there is a real expectation that more than half of the world's top economies in 2050 are likely to be from the 'South' and that they will be collectively larger than the current group of developed countries (HSBC, 2011). They will have a greater role in global trade and investment. New currencies may become more powerful (e.g. the renminbi). Large student and work-forces may mean a better ability to develop eco-, nano-, bio- and info-technologies. New communication devices may mean new threats to information, industrial, technological and national security as secrets leak, knowledge is not filtered for accuracy and intellectual rights are disregarded. As economic, military and information power shifts to other parts of the world, these countries will want to shape the global order including that in the environmental sphere (EC, 2012). They may reproduce lessons from the current recalcitrance of the USA. Those to whom a 'new international economic order' was denied in the 20th century may be less willing to fight for it in the 21st century when and if the tables are turned. Their own vision of global order may also be quite different. While this is a negative argument, positive arguments are that if the rule-of-law is essential in modern states, and needs to be promoted in developing countries by the World Bank and development agencies (Santiso, 2001), then by logical extension it becomes necessary at the international level (Fitzpatrick, 2003; Sandelius, 1933). Law is needed to promote the legitimacy of international action (Anand, 2004; Fitzpatrick, 2003).

10.4.6 *Towards constitutionalization*

The rule of law can be institutionalized through constitutionalization. One can argue that such a process has started already with the UN Charter (Fassbender, 1998) and with the adoption of legal norms in a wide variety of areas from human rights through to environmental issues (Bianchi, 2002: 263). National constitutions include first-, second- and third-generation rights – where second-generation rights focus on subsistence rights and the right to a healthy environment and third-generation rights focus on the rights of individuals and groups including indigenous peoples (Law and Versteeg, 2011). The inclusion of these rights could form a key element of such constitutionalism. Such rights could also emerge from the Rio Declaration on Environment and Development (Rio Declaration, 1992), the Human

Rights arena (e.g. UNGA, 2010a), the right to development (UNGA, 1986) and the emerging norms on the equitable sharing of transboundary resources in the water arena (ILA, 2004; UN Watercourses Convention, 1997). Constitutionalization can also be promoted through the adoption of obligations *erga omnes*: obligations owed to the international community (Simma, 1994). As planetary and social boundaries are crossed, as environmental and human security and rights are threatened, the legal system will have to rise to this challenge. *Erga omnes* norms could draw inspiration from the rights mentioned above and aim to protect these. The constitutionalization process would also include a key element of judicial review. It would not only have to take into account the need for a normative framework at global level, but also that the shift from government to governance implies that the state may not always be the major actor in international relations – so it needs to go beyond creating a normative level playing field for states but also create one for the rules of engagement for non-state actors and individuals.

However, as Koskenniemi (2009: 17) warns us, a commitment to constitutionalization is perhaps just as problematic as streamlining policy to attain imperial preferences. 'The question remains always: what kind of (or whose) law, and what type of (and whose) preference?' (p. 17). Even if constitutionalism aims at transparency and accountability, there is a risk that it becomes a 'facade for statis' and that it leaves out more than it includes. There are others who argue that the rule of law project cannot apply as there is no global government to protect individuals from and the rule of law traditionally protects individuals. If the rule of law is promoted internationally, this may violate the rights of citizens (Waldron, 2011). In the promotion of a discourse on constitutionalization, these challenges can be accounted for. Despite all the critique, Koskenniemi (2009: 17) argues that:

I totally approve of the political move to re-define the managerial world of international institutions through constitutional or administrative vocabularies – not because of the intrinsic worth of those vocabularies, however, but for the critical challenge they pose to today's culture of a-political expert rule, and perhaps for the appeal of the (Kantian) perfectibility that they set up as a regulative goal for human institutions [footnotes deleted].

10.5 Towards the future

If this book has been convincing that an inevitable and necessary element of our common future is a transition to the rule of law at global level, then a key argument is whether and how such a transition can be achieved. The first step is the articulation of an idea and the broad-based acceptance of the idea. Scholars have increasingly articulated and pushed the idea. This idea is also finding support in civil society. While the US government has traditionally not been in favour of this idea, recent polls show that there is growing global support for world order and China,

Germany and the USA had the largest majorities supporting this idea (Council on Foreign Relations, 2011). The South Summit of the G77 and China in 2000 (Art. 6) stated clearly: 'we are committed to a global system based on the rule of law, democracy in decision-making and full respect for the principles of international law and the Charter of the United Nations. The new global system must reflect these principles.'

The second step is the role of actors who push for its acceptance at the global level. A global social movement to this end is needed. The Global Transition 2012, which promotes social support for a global green and fair economy that meets minimum needs, stays within planetary boundaries and is adaptive could possibly be one such social institution. A further step is the role of formal bodies to promote this idea – the European Union that has been promoting the rule of law in new member states is already increasingly positive to this idea (EC, 2012). But there needs to be a coming together of other formal bodies all demanding the same goal.

And finally, the moment needs to be ripe. As the global community faces the quadruple environmental, financial, economic and demographic crises; as Katrina and Sandy reveal how major powers can also be vulnerable to extreme weather events; as Typhoon Haiyan shows how extreme weather events can be and the extent of devastation they can wreak in middle-income countries like the Philippines, as many countries stand on the brink of reinvesting in new forms of fossil fuel; there appears to be a window of opportunity before the balance of power shifts inexorably to the South; this is the right moment to push for the completion of this project!

References

Scientific literature

Abbott, K. W. (1989). Modern international relations theory: A prospectus for international lawyers. *Yale Journal of International Law*, 14(2), 335–411.

Adams, T. B. (2003). Is there a legal future for sustainable development in global warming? Justice, economics and protecting the environment. *Georgetown International Environmental Law Review*, 16, 77–126.

Agarwal A. and Narain, S. (1991). *Global Warming in an Unequal World: A Case of Environmental Colonialism*. New Delhi: Centre for Science and Environment.

Agarwal A. and Narain, S. (1992). *Towards a Green World. Should Global Environmental Management be Built on Legal Conventions or Human Rights?* New Delhi: Centre for Science and Environment.

Agrawal, A., Nepstad, D. and Chhatre, A. (2011). Reducing emissions from deforestation and forest degradation. *Annual Review of Environmental Resources*, 36, 11.1–24.

Akin, J. (2012). Civil justice in the mountains: The Bolivian Andes as grounds for climate reform. *Colorado Journal of International Environmental Law & Policy*, 23, 433–73.

Akumu, G. (1994). Mitigation strategy or hidden agenda. *Tiempo*, 11, 2.

Allen, M. (2003). Liability for climate change: Will it ever be possible to sue anyone for damaging the climate? *Nature*, 421, 891–2.

Allenby, B. (2012). Climate change after the Durban Conference: The dangers of cultural lock-In. *Environmental Quality Management*, 21(4), 19–23.

Alston, P. (1997). The myopia of the handmaidens: International lawyers and globalisation. *European Journal of International Law*, 8(3), 435–48.

Aminzadeh, S. C. (2006–2007). Moral imperative: The human rights implications of climate change, a note. *Hastings International & Comparative Law Review*, 30, 231–66.

Anand, R. (2004). *International Environmental Justice: A North-South Dimension*. Aldershot: Ashgate.

Andonova, L.B. and Alexieva, A. (2012). Continuity and change in Russia's climate negotiations positions and strategy. *Climate Policy*, 12(5), 614–29.

Andresen, S. and Gulbrandsen, L.H. (2003). The role of green NGOs in promoting climate compliance, Report 4/2003. Lysaker: Fridtjof Nansen Institute.

Angel, R. (2006). Feasibility of cooling the earth with a cloud of small spacecraft near the inner Lagrange point (L1). *Proceedings of the National Academy of Sciences*, 103(46), 17184–9.

Angelsen, A., Brown, S., Loisel, C. et al. (2009). *Reducing emissions from deforestation and forest degradation (REDD): An options assessment report.* Meridian Institute.

Arts, K. (2009). A child rights perspective on climate change. In *Climate Change and Sustainable Development: New Challenges for Poverty Reduction*, ed. M. A. Salih. Cheltenham: Edward Elgar, pp. 79–93.

Arts, K. and Gupta, J. (2004). Sustainable development in climate change and hazardous waste law: Implications for the progressive development of international law. In *The Law of Sustainable Development*, ed. N. Schrijver and F. Weiss. Dordrecht: Kluwer Academic Publishers, pp. 519–51.

Ashe, J., Lierop, R. V. and Cherian, A. (1999). The role of the alliance of small island states (AOSIS) in the negotiation of the United Nations Framework Convention on Climate Change (UNFCCC). *Natural Resource Forum*, 23(3), 209–20.

Ausubel, J. (1991). A second look at the impacts of climate change. *American Scientist*, 79, 210–21.

Ayers, J. M. and Huq, S. (2009). Supporting adaptation to climate change: What role for official development assistance? *Development Policy Review*, 27, 675–92.

Azar, C. (1998). Are optimal CO_2 emissions really optimal? Four critical issues for economists in the greenhouse. *Environmental and Resource Economics*, 11(3–4), 301–15.

Azar, C. and Schneider, S. H. (2002). Are the economic costs of stabilizing the atmosphere prohibitive? *Ecological Economics*, 42, 73–80.

Azar, C. and Schneider, S. H. (2003). Are the economic costs of (non-)stabilizing the atmosphere prohibitive? A response to Gerlagh and Papyrakis. *Ecological Economics*, 46, 329–32.

Bachrach, P. and Baratz, M. S. (1970). *Power and Poverty: Theory and Practice.* London/ Oxford: Oxford University Press.

Baer, T., Fieldman, G., Athanasiou, T. and Kartha, S. (2008). Greenhouse development rights: Towards an equitable framework for global climate policy. *Cambridge Review of International Affairs*, 21(4), 649–69.

Bafundo, N. E. (2006). Compliance with the ozone treaty: Weak states and the principle of common but differentiated responsibilities. *American University International Law Review*, 21(3), 461–95.

Balasubramanium, L. (1992). Rootless in the park. *The Illustrated Weekly of India*, 26, 32–4.

Bals, C., Warner, K. and Butzengeiger, S. (2006). Insuring the uninsurable: Design options for a climate change funding mechanism. *Climate Policy*, 6, 637–47.

Banuri, T., Weyant, J., Akumu, G. et al. (2001). Setting the stage: Climate change and sustainable development. In *Climate Change 2001: Mitigation, Report of Working Group III of the Intergovernmental Panel on Climate Change.* Cambridge: Cambridge University Press, pp. 73–114.

Barbier, E. (2011). The policy challenges for green economy and sustainable economic development. *Natural Resources Forum*, 35(3), 233–45.

Barnett, J. (2001). *The Meaning of Environmental Security, Ecological Politics and Policy in the New Security Era.* London: Zed Books.

Barnett, J. (2008). The worst of friends: OPEC and G-77 in the climate regime. *Global Environmental Politics*, 8(4), 1–8.

Barnett, J. and Dessai, S. (2002). Articles 4.8 and 4.9 of the UNFCCC: Adverse effects and the impacts of response measures. *Climate Policy*, 2(2), 231–9.

Barrett, S. (2002). Consensus treaties. *Journal of Institutional and Theoretical Economics*, 158, 529–47.

Barrett, S. (2009). The coming global climate technology revolution. *Journal of Economic Perspectives*, 23(2), 53–75.

Baum, R. (1986). Modernization and legal reform in post-Mao China: The rebirth of socialist legality. *Studies in Comparative Communism*, 19(2), 69–103.

Benedick, R. E. (1993). Perspectives of a negotiation practitioner. In *International Environment Negotiation*, ed. G. Sjostedt. Laxenberg: IIASA, pp. 219–43.

Benford, G. (1997). Climate controls. If we treated global warming as a technical problem instead of a moral outrage, we could cool the world. *Reason Online* (accessed 13 May, 2013 from http://reason.com/archives/1997/11/01/climate-controls).

Berkhout, F. (1998). Aggregate resource efficiency. In *Managing a Material World: Reflections on Industrial Ecology*, ed. P. Vellinga, F. Berkhout and J. Gupta. *Environment and Policy Series*. Dordrecht: Kluwer Academic Publishers, pp. 165–90.

Betzold, C., Castro, P. and Weiler, F. (2012). AOSIS in the UNFCCC negotiations: From unity to fragmentation? *Climate Policy*, 12(5), 591–61.

Bhole, L. M. (1992). Environmental protection through sustainable social order. *The Indian Journal of Social Science*, 5(3), 289–99.

Bianchi, A. (2002). Ad-hocism and the rule of law. *European Journal of International Law*, 13(1), 263–72.

Biermann, F., Abbott, K., Andresen, S. et al. (2012a). Navigating the anthropocene: Improving earth system governance. *Science*, 335(6074), 1306–7.

Biermann, F., Abbott, K., Andresen, S. et al. (2012b). Towards a strengthened institutional framework for global sustainability: Key insights for social science research. *Current Opinion in Environmental Sustainability*, 4(1), 51–60.

Bloom, N. (2009). The impact of uncertainty shocks. *Econometrica*, 77(3), 623–85.

Bodansky, D. (1993). The United Nations Framework Convention on Climate Change: A commentary. *Yale Journal of International Law*, 18, 451–588.

Bodansky, D. (2001). Bonn voyage: Kyoto's uncertain revival. *The National Interest*, Fall 2001, 45–55.

Bond, P. (2011). From Copenhagen to Cancún to Durban: Moving deckchairs on the climate Titanic. *Capitalism Nature Socialism*, 22(2), 3–26.

Botchway, F. N. (2001). Good governance: The old, the new, the principle, and the elements. *Florida Journal of International Law*, 13(2), 159–210.

Bouwer, L. M. (2011). Have disaster losses increased due to anthropogenic climate change? *Bulletin of the American Meteorological Society*, 92(1), 39–46.

Bowen, M. (2008). *Censoring Science: Inside the Political Attack on Dr. James Hansen and the Truth of Global Warming*. New York: Dutton.

Boyd, E. (2009). Governing the Clean Development Mechanism: Global rhetoric versus local realities in carbon sequestration projects. *Environment and Planning A*, 41(10), 2380–95.

Boyd, E., Corbera, E. and Estrada, M. (2008). UNFCCC negotiations (pre-Kyoto to COP-9): What the process says about the politics of CDM-sinks. *International Environmental Agreements: Politics, Law and Economics*, 8, 95–112.

Boyd, E., Hultman, N., Timmons Roberts, J., et al. (2009). Reforming the CDM for sustainable development: Lessons learned and policy futures. *Environmental Science & Policy*, 12, 820–31.

Boykoff, M. T. and Boykoff, J. M. (2004). Balance as bias: global warming and the US prestige press. *Global Environmental Change*, 14, 125–36.

Boyle, F. (1980). The irrelevance of international law: The schism between international law and international politics. *California Western International Law Journal*, 10(2), 193–219.

Brand, U. (2012). Green economy – the next oxymoron? No lessons learned from failures of implementing sustainable development. *GAIA*, 21(1), 28–32.

Bratspeis, R. M. (2011–2012). Human rights and environmental regulation. *N. Y. U. Environmental Law Journal*, 19, 225–302.

Braybrooke, D. and Lindblom, C. E. (1963). *A Strategy of Decision: Policy Evaluation as a Social Process*. New York: The Free Press.

Brooks, N., Gash, J., Hulme, M. et al. (2005). *Climate Stabilisation and 'Dangerous' Climate Change: A Review of the Relevant Issues. Scoping Study*. London: DEFRA.

Brown, L. (1981). *Building a Sustainable Society*. Washington, DC: Worldwatch Institute.

Brown, O. (2008). *Migration and Climate Change*. IOM Migration Research Series No.31. Geneva: International Organization for Migration.

Bruce, J. P. (1996). In search of cost-effective mitigation options. *United Nations Climate Change Bulletin*, 12(3).

Brunnée, J. (2002). COPing with consent: Law-making under multilateral environmental agreements. *Leiden Journal of International Law*, 15(1), 1–52.

Bulkeley, H. and Betsill, M. (2003). *Cities and Climate Change: Urban Sustainability and Global Environmental Governance*. London and New York: Routledge Studies in Physical Geography and the Environment.

Bulkeley, H., Andonova, L., Bäckstrand, K. et al. (2012). Governing climate change transnationally: Assessing the evidence from a database of sixty initiatives. *Environment and Planning C: Government and Policy*, 30, 591–612.

Buob, S. and Stephan, G. (2011). To mitigate or to adapt: How to confront global climate change. *European Journal of Political Economy*, 27, 1–16.

Burns, W. G. C. (2004). The exigencies that drive potential causes of action for climate change damages at the international level. *American Society of International Law Proceedings*, 98, 223–7.

Bush, G. W. (2001). President Bush Discusses Global Climate Change, June 11 (accessed from www.whitehouse.gov/news/releases/2001/06/20010611–2.html).

Cairo Compact (1989). *The Cairo Compact: Towards a Concerted Worldwide Response to the Climate Crises*. Cairo: World Conference on Preparing for Climate Change.

Cameron, E. and Limon, M. (2012). Restoring the climate by realizing rights: The role of the international human rights system. *Review of European Community and International Environmental Law*, 21(3), 204–19.

Caney, S. (2008). Human rights, climate change, and discounting. *Environmental Politics*, 17(4), 536–55.

Cao, J. (2010). Reconciling human development and climate protection: A multistage hybrid climate policy architecture. In: *Post-Kyoto International Climate Policy: Implementing Architectures for Agreement*, ed. J. E. Aldy and R. N. Stavins. Cambridge: Cambridge University Press, pp. 563–98.

Carbon Disclosure Project (2011a). *CDP Global 500 Report 2011: Accelerating Low Carbon Growth* (accessed from www.cdproject.net/CDPResults/CDP-G500–2011-Report.pdf).

Carbon Disclosure Project (2011b). *CDP S&P 500 Report 2011: Strategic Advantage through Climate Change Action*. London: Carbon Disclosure Project (accessed 17 May 2013 from www.cdproject.net/CDPResults/CDP-2011-SP500.pdf).

Carbon Disclosure Project (2013). *Sector Insights: What is Driving Climate Change Action in the World's Largest Companies?* Global 500 Climate Change Report, 2013.

Caviglia-Harris, J. L., Chambers, D. and Kahn, J. R. (2009). Taking the 'U' out of Kuznets: A comprehensive analysis of the EKC and environmental degradation. *Ecological Economics*, 68(4), 1149–59.

Chakravarty, S., Chikkatur, A., de Coninck, H. et al. (2009). Sharing global CO_2 emission reductions among one billion high emitters. *Proceedings of the National Academy of Sciences*, 196(29), 11884–8.

Charlson, R. J., Lovelock, J. E., Andreae, M. O. and Warren, S. G. (1987). Oceanic phyto-plankton, atmospheric sulphur, cloud albedo and climate. *Nature*, 326, 655–61.

Chimni, B. S. (2006). Third World Approaches to international law: A manifesto. *International Community Law Review*, 8(1), 3–28.

Clapp, J. (2009). Food price volatility and vulnerability in the Global South: Considering the global economic context. *Third World Quarterly*, 30, 1183–96.

Clements, T. (2010). Reduced expectations: The political and institutional challenges of REDD+. *Oryx*, 44(3), 309–10.

Commoner, B. (1971). *The Closing Circle: Nature, Man and Technology*. New York: Knopf.

Copeland, B. R. and Taylor, M. S. (2004). Trade, growth, and the environment. *Journal of Economic Literature*, 42(1), 7–71.

Correll, H. (2001). The visible college of international law: Towards the rule of law in inter-national relations. *American Society of International Law Proceedings*, 95, 262–70.

Cortez, R., Saines, R., Griscom, B. et al. (2010). *A Nested Approach to REDD+: Structuring Effective and Transparent Incentive Mechanisms for REDD+ Implementation at Multiple Scales*. Washington, DC: Nature Conservancy.

Council on Foreign Relations (2011). *Public Opinion on Global Issues* (accessed 14 May 2013 from www.cfr.org/thinktank/iigg/pop/).

Courtois, P. and Haeringer, G. (2011). Environmental cooperation: Ratifying second-best agreements. *Public Choice*, 15, 20.

Craig, P. (1997). Formal and substantive conceptions of the rule of law: An analytical framework. *Public Law*, 3, 467–87.

Crowley, J. (2012). Climate change, climate knowledge and human rights. *Presentation at the OHCHR Seminar to Address the Adverse Impacts of Climate Change on the Full Enjoyment of Human Rights*. Geneva, 23–4 February 2011, UNESCO Division of Ethics, Science and Society.

Crutzen, P. J. (2006). Albedo enhancement by stratospheric sulphur injections: A contribu-tion to resolve a policy dilemma? *Climatic Change*, 77, 211–19.

Cunningham, W. P. and Cunningham, M. A. (2002). *Principles of Environmental Science: Inquiry and applications*. Boston: McGraw-Hill.

Cutlip, L. and Fath, B. D. (2012). Relationship between carbon emissions and eco-nomic development: Case study of six countries. *Environment, Development, and Sustainability*, 14(3), 433–53.

Dahl, A. (2000). Competence and subsidiarity. In *Climate Change and European Leadership: A Sustainable Role for Europe*, ed. J. Gupta and M. Grubb. Dordrecht: Kluwer Academic Publishers, pp. 203–20.

Dannenberg, A., Löschel, A., Paolacci, G., Reif, C. and Tavoni, A. (2011). *Coordination under Threshold Uncertainty in a Public Goods Game*. Mannheim: Centre for European Economic Research (accessed 15 May 2013 from http://ftp.zew.de/pub/zewdocs/dp/dp11065.pdf).

Dasgupta, C. (1994). The climate change negotiations. In *Negotiating Climate Change: The Inside Story of the Rio Convention*, ed. I. M. Mintzer and J. A. Leonard. Cambridge: Cambridge University Press, pp. 129–48.

Dasgupta, P. (1993). *An Inquiry into Well-being and Destitution*. Oxford: Clarendon Press.

Davis, J. and Tan, C. (2010). *Tackling Climate Change through Adaptation Finance in the Least Developed Countries: Is the LDC Fund Still Fit for Purpose?* Geneva: UNCTAD.

Davis, W. J. (1996). The Alliance of Small Island States (AOSIS): The international con-science. *Asia-Pacific Magazine*, 2, 17–22.

De Bruyn, S. (1998). Dematerialization and rematerialization. In *Managing a Material World: Reflections on Industrial Ecology*, ed. P. Vellinga, F. Berkhout and J. Gupta. *Environment and Policy Series*. Dordrecht: Kluwer Academic Publishers, pp. 147–64.

De Bruyn, S. M. and Opschoor, J. B. (1997). Developments in the throughput–income relationship: Theoretical and empirical observations. *Ecological Economics*, 20(3), 255–69.

De Gryze, S. and Durschinger, L. (2010). *An Integrated REDD Offset Program (IREDD) for Nesting Projects under Jurisdictional Accounting*. San Francisco: Terra Global Capital LLC.

De Larragán, J. de C. (2011). Liability of Member States and the EU in view of the international climate change framework: Between solidarity and responsibility. In *Climate Change Liability*, ed. M. Faure and M. Peeters. Cheltenham: Edward Elgar, pp. 55–89.

Dellink, R. and Finus, M. (2009). Uncertainty and climate treaties: Does ignorance pay? *Stirling Economics Discussion Papers, 2009–2015* (accessed 15 May 2013 from http://hdl.handle.net/1893/1476).

Den Elzen, M. G. J. and Höhne, N. (2010). Sharing the reduction effort to limit global warming to 2°C. *Climate Policy*, 10, 247–60.

Denton, F. (2010). Financing adaptation in least developed countries in West Africa: Is finance the 'real deal'? *Climate Policy*, 10, 655–71.

Depledge, J. (2000). Tracing the origins of the Kyoto Protocol: An article-by-article textual history. UN Doc. FCCC/TP/2000/2. 25 November, 2000.

Depledge, J. (2006). The opposite of learning: Ossification in the climate change regime. *Global Environmental Politics*, 6(1), 1–22.

Dessai, S., Adger, W. N., Hulme, M. et al. (2003). Defining and experiencing dangerous climate change. *Climatic Change*, 64(1), 11–25.

Dinda, S. (2004). Environmental Kuznets curve hypothesis: A survey. *Ecological Economics*, 49(4), 431–55.

Dlugolecki, A. (2008). Climate change and the insurance sector. *The Geneva Papers*, 33, 71–90.

Doelle, M. (2004). Linking the Kyoto Protocol and other multilateral environmental agreements: From fragmentation to integration? *Journal of Environmental Law and Practice*, 14, 75–104.

Douthwaite, R. (1995). Who says that life is cheap? *The Guardian*, 1 November, 1995. London: The Environment Section.

Dovers, S. and Handmer, J. (1993). Contradictions in sustainability. *Environmental Conservation*, 20(3), 217–22.

Dworkin, R. (1985). *A Matter of Principle*. Cambridge: Harvard University Press.

Easterly, W. (2007). Was development assistance a mistake? *American Economic Review*, 97(2), 328–32.

Ebert, U. and Welsch, H. (2011). Optimal response functions in global pollution problems can be upward sloping: Accounting for adaptation. *Environmental Economics and Policy Studies*, 13, 129–38.

Editorial (2011). The mask slips. *Nature*, 480, 292.

Eliasch, J. (2008). *Climate Change: Financing Global Forests*. London: Office of Climate Change.

Emsley, J. (ed.) (1996). *The Global Warming Debate: The Report of the European Science Forum*. Dorset: Bournemouth Press Limited.

Esquith, S. L. (1999). Toward a democratic rule of law: East and West. *Political Theory*, 27(3), 334–56.

Fankcuser, S., Smith, J. B. and Tol, R. S. J. (1999). Weathering climate change: Some simple rules to guide adaptation decisions. *Ecological Economics*, 30, 67–78.

Fassbender, B. (1998). The United Nations Charter as Constitution of the International Community. *Columbia Journal of Transnational Law*, 36(3), 529–619.

Faure, M. and Peeters, M. (eds) (2011). *Climate Change Liability*. Cheltenham: Edward Elgar.

FCPF (2011). *Forest Carbon Partnership Facility 2011 Annual Report*. Washington, DC: World Bank.

Finus, M. and Pintassilgo, P. (2011). International environmental agreements under uncertainty: Does the 'veil of uncertainty' help? *Oxford Economic Papers*, 64(4), 736–64.

Fischetti, M. (2005). They saw it coming. *The New York Times*, 2 September (accessed from www.nytimes.com/2005/09/02/opinion/02fischetti.html?pagewanted=all&_r=0).

Fitzpatrick, P. (2003). 'Gods would be needed…': American empire and the rule of (international) law. *Leiden Journal of International Law*, 16(3), 429–66.

Fletcher, A. L. (2009). Clearing the air: The contribution of frame analysis to understanding climate policy in the United States. *Environmental Politics*, 18(5), 800–16.

Foqué, R. (1998). Global governance and the rule of law. In *International Law: Theory and Practice, Essays in Honour of Eric Suy*, ed. K. Wellens. The Hague: Kluwer Law International, pp. 225–44.

Forsyth, T. (2005). Enhancing climate technology transfer through greater public-private cooperation: Lessons from Thailand and the Philippines. *Natural Resources Forum*, 29, 165–76.

Franck, T. M. (1995). *Fairness in International Law and Institutions*. Oxford: Clarendon Press.

Frankel, J. (2010). An elaborated proposal for a global climate policy architecture: Specific formulas and emission targets for all countries in all decades. In *Post-Kyoto International Climate Policy: Implementing Architectures for Agreement: Research from the Harvard Project on International Climate Agreements*, ed. J. E. Aldy and R. N. Stavins. Cambridge: Cambridge University Press, pp. 31–87.

Freestone, D. and Rayfuse, R. (2008). Iron ocean fertilization and international law. *Marine Ecology Progress Series*, 364, 227–33.

French, D. (2000). Developing states and international environmental law: The importance of differentiated responsibilities. *International and Comparative Law Quarterly*, 49(1), 35–60.

Fry, I. (2008). Reducing emissions from deforestation and forest degradation: Opportunities and pitfalls in developing a new legal regime. *Review of European Community and International Environmental Law*, 17(2), 166–82.

Fuhr, L., Schalatek, L. and Omari, K. (2011). *COP 17 in Durban: A Largely Empty Package*. Berlin: Heinrich Boll Stiftung, The Green Political Foundation.

Fuller, L. L. (1964). *The Morality of Law*. New Haven: Yale University Press.

Functovicz, S., O'Connor, M. and Ravetz, J. (1996). Emergent complexity and ecological economics. In *Economy and Ecosystems in Change*, ed. J. van der Straten and J. van den Bergh. Washington, DC: Island Press, pp. 75–95.

G77 and China (2008). *Financial Mechanisms for Meeting Financial Commitments under the Convention* (25 August). Bonn: Secretariat to the UNFCCC.

Garcia-Amador, F. V. (1990). *The Emerging International Law of Development: A New Dimension of International Economic Law*. New York: Oceana Publications.

Garnaud, B. (2009). An Analysis of Adaptation Negotiations in Poznan, IDDRI, Policy brief No. 01/2009.

Gaskill, A. (2004). Summary of meeting with US DOE to discuss geoengineering options to prevent abrupt and long-term climate change (accessed 13 November 2009 from www.see.ed.ac.uk/~shs/Climate%20change/Geo-politics/Gaskill%20DOE.pdf).

Georgiev, D. (1993). Politics or rule of law: Deconstruction and legitimacy in international law. *European Journal of International Law*, 4, 1–14.

Gerlach, L. P. (1992). Problems and prospects of institutionalising ecological interdependence in a world of local independence. In *Global Warming and the Challenge of International Cooperation: An interdisciplinary assessment*, ed. G. Bryner. Utah: Brigham Young University, pp. 69–86.

Gillespie, A. (2004). Small island states in the face of climate change: The end of the line in international environmental responsibility. *UCLA Journal of Environmental Law and Policy*, 22, 107–29.

Ginther, K. and De Waart, P. J. I. M. (1995). Sustainable development as a matter of good governance: An introductory view. In *Sustainable Development and Good Governance*, ed. K. Ginther, E. Denters and P. J. I. M. De Waart. Dordrecht: Kluwer Academic Publishers, pp. 1–14.

GMI (2012). *The US Government's Global Methane Initiative Accomplishments* (accessed 13 May 2013 from www.cpa.gov/globalmethane/pdf/2012-accomplish-report/GMI_USG2012_Full-Report.pdf).

Gore, A. (1997). Briefing on Kyoto Warming Pact, 11 December 1997.

Gouritin, A. (2011). Potential liability of European states under the ECHR for failure to take appropriate measures with a view to adaptation to climate change. In *Climate Change Liability*, ed. M. Faure and M. Peeters. Cheltenham: Edward Elgar, pp. 134–64.

Government of Argentina (2008). *Argentina's Views on Enabling the Full, Effective, And Sustained Implementation of the Convention through Long-Term Cooperative Action Now, Up To, and Beyond 2012*. November (accessed 13 May 2013 from http://unfccc.int/files/kyoto_protocol/application/pdf/argentinabap300908.pdf).

Government of China (2008). *China's Views on enabling the full, effective and sustained implementation of the Convention through long-term cooperative action now, up to and beyond 2012*. 28 September (accessed 13 May 2013 from http://unfccc.int/files/kyoto_protocol/application/pdf/china_bap_280908.pdf).

Government of India (2008). *Government of India Submission on Financing Architecture for Meeting Financial Commitments Under The UNFCCC* (accessed 13 May 2013 from http://unfccc.int/files/kyoto_protocol/application/pdf/indiafinancialarchitecture171008.pdf).

Grainger, A. and Obersteiner, M. (2011). A framework for structuring the global forest monitoring landscape in the REDD+ era. *Environmental Science & Policy*, 14(2), 127–39.

Greene, O. (1996). Lessons from other international environmental agreements. In *Sharing the Effort: Options for Differentiating Commitments on Climate Change*, ed. M. Paterson and M. Grubb. London: The Royal Institute of International Affairs, pp. 23–44.

Greenpeace (1998). *Greenpeace Analysis of the Kyoto Protocol*, Greenpeace Briefing Paper. Amsterdam: Greenpeace International.

Grey, K. and Gupta, J. (2003). The United Nations climate change regime and Africa. In *International Environmental Law and Policy in Africa*, ed. K. R. Gray and B. Chaytor. Dordrecht: Kluwer Academic Publishers, pp. 60–81.

Griffiths, T. (2008). *Seeing 'REDD'? Forests, Climate Change Mitigation and the Rights of Indigenous Peoples. Update for Poznan (COP 14)*. Moreton-in-Marsh: Forest Peoples Programme.

Grossman, D. A. (2003). Warming up to a not-so-radical idea: Tort-based climate change litigation. *Colombia Journal of Environmental Law*, 28, 1–61.

Grossman, G. M. (1995). Pollution and growth: what do we know? In *The Economics of Sustainable Development*, ed. I. Goldin and L. A. Winters. Cambridge: Cambridge University Press, pp. 19–42.

Grossman, G. M. and Krueger, A. B. (1995). Economic growth and the environment. *Quarterly Journal of Economics*, 110(2), 353–77.

Group of 77 South Summit (2000). *Declaration of the South Summit*, Havana, 10–14 April 2000.

Grubb, M. and Gupta, J. (2000). Leadership. In *Climate Change and European Leadership: A Sustainable Role for Europe*, ed. J. Gupta and M. Grubb. Dordrecht: Kluwer Academic Publishers, pp. 15–24.

Grubb, M. and Yamin, F. (2001). Climatic collapse at The Hague: What happened, why, and where do we go from here? *International Affairs*, 77(2), 261–76.

Gupta, J. (1995). The global environment facility in its North–South context. *Environmental Politics*, 4(1), 19–43.

Gupta, J. (1997). *The Climate Change Convention and Developing Countries: From Conflict to Consensus?* Dordrecht: Kluwer Academic Publishers.

Gupta, J. (1998). Leadership in the climate regime: Inspiring the commitment of developing countries in the post-Kyoto phase. *Review of European Community and International Environmental Law*, 7(2), 178–88.

Gupta, J. (2000). *On Behalf of My Delegation: A Guide for Developing Country Climate Negotiators*. Washington, DC: Center for Sustainable Development of the Americas.

Gupta, J. (2003). Engaging developing countries in climate change: KISS and make-up! In *Climate Policy for the 21st Century: Meeting the Long-Term Challenge of Global Warming*, ed. D. Michel. Washington, DC: Center for Transatlantic Relations, Johns Hopkins University Press, pp. 233–64.

Gupta, J. (2005). L'Union Européene, Leader de la Politique Internationale du Changement Climatique? In *L'Union Européenne, Acteur Internationale*, ed. D. Helly and F. Petiteville. Paris : L'Harmattan, pp. 253–66.

Gupta, J. (2006). Increasing disenfranchisement of developing country negotiators in a multi-speed world. In *The Politics of Participation in Sustainable Development Governance*, ed. J. F. Green and W. B. Chambers. Tokyo: United Nations University Press, pp. 21–39.

Gupta, J. (2007). Legal steps outside the Climate Convention: Litigation as a tool to address climate change. *Review of European Community and International Environmental Law*, 16(1), 76–86.

Gupta, J. (2009). Climate change, development and evaluation: Can flexibility instruments promote sustainable development? In *Evaluating Climate Change and Development*, ed. R. van den Berg and O. Feinstein. New Brunswick/New Jersey: Transaction Publishers, pp. 47–64.

Gupta, J. (2010a). Global governance: Development cooperation. In *Mainstreaming Climate Change in Development Cooperation: Theory, Practice and Implications for the European Union*, ed. J. Gupta and N. van der Grijp. Cambridge: Cambridge University Press, pp. 99–133.

Gupta, J. (2010b). Mainstreaming climate change: A theoretical exploration. In *Mainstreaming Climate Change in Development Cooperation: Theory, Practice and Implications for the European Union*, ed. J. Gupta and N. van de Grijp. Cambridge: Cambridge University Press, pp. 67–96.

Gupta, J. (2010c). A history of international climate change policy. *Wiley Interdisciplinary Reviews: Climate Change*, 1(5), 636–53.

Gupta, J. (2011a). Developing countries: Trapped in the web of sustainable development governance. In *Transnational Administrative Rule-Making: Performance, Legal Effects and Legitimacy*, ed. O. Dilling, M. Herberg and G. Winter. Oxford/Portland: Hart Publishing, pp. 305–30.

Gupta, J. (2011b). Climate change: A GAP analysis based on Third World approaches to international law. In *German Yearbook of International Law*, 53 (2010) but published 2011. Berlin: Duncker & Humblot, pp. 341–70.

Gupta, J. (2012a). Glocal forest and REDD+ governance: Win-win or lose-lose? *Current Opinion in Environmental Sustainability*, 4, 1–8.

Gupta, J. (2012b). Negotiating challenges and climate change. *Climate Policy*, 12(5), 630–44.

Gupta, J. and Hisschemöller, M. (1997). Issue-linkages: A global strategy towards sustainable development. *International Environmental Affairs*, 9(4), 289–308.

Gupta, J. and Ringius, L. (2001). The EU's climate leadership: Between ambition and reality. *International Environmental Agreements: Politics, Law and Economics*, 1(2), 281–99.

Gupta, J. and Sanchez, N. (2012a). Global green governance: The green economy needs to be embedded in a global green and equitable rule of law polity. *Review of European Community and International Environmental Law*, 21(1), 12–22.

Gupta, J. and Sanchez, N. (2012b). Elaborating the common but differentiated responsibilities principle: Experiences in the WTO. Legal working paper. Rome: International Developmental Law Organization.

Gupta, J. and Sanchez, N. (2013). The CBDR Principle elaborated in relation to other principles of law. In *The Global Community Yearbook of International Law and Jurisprudence, Global Trends: Law, Policy & Justice. Essays in Honour of Professor Giuliana Ziccardi Capaldo*, ed. M. C. Bassiouni, G. Joanna, P. Mengozzi et al. Oxford: Oxford University Press, pp. 23–39.

Gupta, J. and van Asselt, H. (2006). Helping operationalise Article 2: A transdisciplinary methodological tool for evaluating when climate change is dangerous. *Global Environmental Change*, 16(1), 83–94.

Gupta, J. and van Asselt, H. (2007). Towards a fair and equitable framework for post-2012 climate change policy. ICCO policy brief. Utrecht: Interkerkelijke Organisatie voor Ontwikkelingssamenwerking (ICCO).

Gupta, J. and van der Grijp, N. (2000). Perceptions of the EU's role. In *Climate Change and European Leadership: A Sustainable Role for Europe*, ed. J. Gupta and M. Grubb. Dordrecht: Kluwer Academic Publishers, pp. 63–82.

Gupta, J. and van der Grijp, N. (2010). Prospects for mainstreaming climate change in development cooperation. In *Mainstreaming Climate Change in Development Cooperation: Theory, Practice and Implications for the European Union*, eds. J. Gupta and N. van der Grijp. Cambridge University Press.

Gupta, J., Ahlers, R. and Ahmed, L. (2010b). The human right to water: Moving toward consensus in a fragmented world. *Review of European Community and International Environmental Law*, 19(3), 294–305.

Gupta, J., Lasage, R. and Stam, T. (2007). National efforts to enhance local climate policy in the Netherlands. *Special Issue of Environmental Sciences*, 4(3), 171–82.

Gupta, J., Opschoor, J. B., Kok, M. (2013). New Paths to International Environmental Cooperation. Advisory Council on International Affairs. Ministry of Foreign Affairs, The Netherlands.

Gupta, J., Persson, Å., Olsson, L. et al. (2010c). Mainstreaming climate change in development cooperation: Conditions for success. In *Adaptation and Mitigation of Climate Change*, ed. M. Hulme and H. Neufeldt. Cambridge: Cambridge University Press, pp. 319–39.

Gupta, J., Termeer, K., Klostermann, J. et al. (2010a). Institutions for climate change: A method to assess the inherent characteristics of institutions to enable the adaptive capacity of society. *Environmental Science and Policy*, 13(6), 459–71.

Gupta, J., van Asselt, H., Weber, D. et al. (2003). Helping perationalise Article Two (HOT): Report of Phase 1 of a science-based policy dialogue on fair and effective ways to avoid dangerous interference with the climate system and implications for post-Kyoto policies. IVM report W-03/26. Amsterdam: Institute for Environmental Studies, Vrije Universiteit.

Gupta, J., van der Grijp, N. and Kuik, O. (eds.) (2013). *Climate Change, Forests and REDD: Lessons for Institutional Design*. Oxford/New York: Routledge.

Gupta, S., Tirpak, D., Boncheva, A. I. et al. (2007). Policies, instruments and cooperative arrangements. In *IPCC-3 Fourth Assessment Report*. Cambridge: Cambridge University Press, pp. 747–807.

Guston, D. (2001). Towards a 'best practice' of constructing 'serviceable truths'. In *Knowledge, Power and Participation in Environmental Policy Analysis*, ed. M. Hisschemöller, R. Hoppe, W. Dunn and J. Ravetz. New Brunswick: Transaction Publishers, pp. 97–119.

Haas, P. M. (1989). Do regimes matter? Epistemic communities and Mediterranean pollution control. *International Organization*, 43(3), 377–403.

Hager, B. (2000). *The Rule of Law, A Lexicon for Policy Makers*. Mamsfield: The Mamsfield Center for Pacific Affairs.

Haites, E., Duan, M. S. and Seres, S. (2006). Technology transfer by CDM projects. *Climate Policy*, 6, 327–44.

Hamwey, R. M. (2007). Active amplification of the terrestrial albedo to mitigate climate change: An exploratory study. *Mitigation and Adaptation Strategies for Global Change*, 12, 419–39.

Hancock, E. E. (2005). Red dawn, blue thunder, purple rain: Corporate risk of liability for global climate change and the SEC disclosure dilemma. *Georgetown International Environmental Law Review*, 17(2), 223–51.

Hardin, G. (1968). The Tragedy of the Commons. *Science*, 162(3859), 1243–8.

Hare, W. and Meinshausen, M. (2004). How much warming are we committed to and how much can be avoided? PIK Report No. 93. Berlin: Potsdam Institute for Climate Impact Research.

Haritz, M. (2010). An inconvenient deliberation: The precautionary principle's contribution to the uncertainties surrounding climate change liability. PhD thesis at Maastricht University. Oisterwijk: Box Press Publishers.

Harris, P. (2001). *International Equity and Global Environmental Politics*. London: Ashgate.

Harrop, S. R. and Pritchard, D. J. (2011). A hard instrument goes soft: The implications of the Convention on Biological Diversity's current trajectory. *Global Environmental Change*, 21(2), 474–80.

Haug, C. and Gupta, J. (2013). The emergence of REDD on the global policy agenda. In *Climate Change, Forests and REDD: Lessons for Institutional Design*, ed. J. Gupta, N. van der Grijp and O. Kuik. Oxford/New York: Routledge, pp. 77–98.

Henkin, L. (1969). International organization and the rule of law. *International Organization*, 23(3), 656–82.

Herweijer, C., Ranger, N. and Ward, R. E. T. (2009). Adaptation to climate change: Threats and opportunities for the insurance industry. *The Geneva Papers*, 34, 360–80.

Hey, E. (2000). The climate change regime: An enviro-economic problem and international administrative law in the making. *International Environmental Agreements: Politics, Law and Economics*, 1(1), 75–100.

Hisschemöller, M. (1993). *De Democratie van Problemen, De Relatie Tussen de Inhoud van Beleidsproblemen en Methoden van Politieke Besluitvorming.* Amsterdam: VU Uitgeverij.

Hisschemöller, M. and Gupta, J. (1999). Problem-solving through international environmental agreements: The issue of regime effectiveness. *International Political Science Review*, 20(2), 153–76.

Hoeppe, P. and Gurenko, E. N. (2006). Scientific and economic rationales for innovative climate insurance solutions. *Climate Policy*, 6(6), 607–20.

Honkonen, T. (2009). The principle of common but differentiated responsibilities in post-2012 climate negotiations. *Review of European Community and International Environmental Law*, 18(3), 257–9.

Houghton, J. T., Jenkins, G. J. and Ephraums, J. J. (1990). *Climate Change: The IPCC Scientific Assessment.* Cambridge: Cambridge University Press.

Houghton, J. T., Meira Filho, L. G., Bruce, J. et al. (eds.) (1995). *Climate Change 1994: Radiative Forcing of Climate Change and an Evaluation of the IPCC IS92 Emission Scenarios.* Cambridge: Cambridge University Press.

HSBC (2011). *The World in 2050: Quantifying the Shift in the Global Economy* (accessed 13 May 2013 from www.hsbc.com/~/media/HSBC-com/about-hsbc/in-the-future/pdfs/120508-the-world-in-2050.ashx).

Huq, S., Reid, H., Konate, M. et al. (2004). Mainstreaming adaptation to climate change in least developed countries (LDCs). *Climate Policy*, 4, 25–43.

Idso, S. (1996). Plant responses to rising levels of atmospheric carbon dioxide. In *The Global Warming Debate: The Report of the European Science Forum*, ed. J. Emsley. Dorset: Bournemouth Press Limited, pp. 28–33.

IEA (2011). *World Energy Outlook.* Paris: IEA.

ILA (2004). *Berlin Rules.* Report of the Seventy-First Conference, Berlin. London: ILA.

Ingham A., Ma, J. and Ulph, A. (2005). *Can Adaptation and Mitigation be Complements?* Norwich: Tyndall Centre for Climate Change Research.

IPCC (1997). *The Regional Impacts of Climate Change: An Assessment of Vulnerability.* IPCC Special Report. Cambridge: Cambridge University Press.

IPCC (1999). *Aviation and the Global Atmosphere.* Prepared in collaboration with the Scientific Assessment Panel to the Montreal Protocol on Substances that Deplete the Ozone Layer. Cambridge: Cambridge University Press.

IPCC (2000a). *Methodological and Technological Issues in Technology Transfer. Summary for Policymakers.* IPCC Special Report. Geneva: IPCC.

IPCC (2000b). *Land Use, Land-Use Change, and Forestry.* Cambridge: Cambridge University Press.

IPCC (2003). *Report of the Twentieth Session of the Intergovernmental Panel on Climate Change (IPCC).* Paris, 19–21 February 2003.

IPCC (2004). IPCC Expert Meeting on The Science to Address UNFCCC Article 2 including Key Vulnerabilities. Short report.

IPCC (2005). *Carbon Dioxide Capture and Storage.* Cambridge: Cambridge University Press.

IPCC (2007). *Climate Change 2007, Synthesis Report.* Cambridge: Cambridge University Press.

IPCC (2012). *Managing the Risks of Extreme Events and Disasters to Advance Climate Change Adaptation. A Special Report of Working Groups I and II of the Intergovernmental Panel on Climate Change.* Cambridge/New York: Cambridge University Press.

IPCC-I (1990). *Climate Change: The IPCC Scientific Assessment.* Cambridge: Cambridge University Press.

IPCC-I (1996). *Climate Change 1995: The Science of Climate Change. Contribution of Working Group I to the Second Assessment Report.* Cambridge: Cambridge University Press.

IPCC-I (2001). *Climate Change 2001: The Scientific Basis. Contribution of Working Group I to the Third Assessment Report of the Intergovernmental Panel on Climate Change.* Cambridge: Cambridge University Press.

IPCC-I (2007). *Climate Change 2007: The Physical Science Basis. Contribution of Working Group I to the Fourth Assessment Report of the Intergovernmental Panel on Climate Change.* Cambridge: Cambridge University Press.

IPCC-II (1990). *Climate Change: The IPCC Impacts Assessment.* Canberra: Australian Government Publishing Service.

IPCC-II (1996). *Climate Change 1995: Impacts, Adaptations, and Mitigation of Climate Change. Scientific-Technical Analyses. Contribution of Working Group II to the Second Assessment Report.* Cambridge: Cambridge University Press.

IPCC-II (2007). *Climate Change 2007: Impacts, Adaptation, and Vulnerability. Contribution of Working Group II to the Fourth Assessment Report of the Intergovernmental Panel on Climate Change.* Cambridge: Cambridge University Press.

IPCC-III (1990). *Climate Change: The IPCC Response Strategies.* Washington, DC: Island Press.

IPCC-III (1996). *Economic and Social Dimensions of Climate Change. Contribution of Working Group III to the Second Assessment Report.* Cambridge: Cambridge University Press.

IPCC-III (2001). *Climate Change 2001: Mitigation.* Cambridge: Cambridge University Press.

IPCC-III (2007). *Climate Change 2007: Mitigation of Climate Change. Contribution of Working Group III to the Fourth Assessment Report of the Intergovernmental Panel on Climate Change.* Cambridge: Cambridge University Press.

IUCN (2010) *Indigenous Peoples and REDD-Plus: Challenges and Opportunities for the Engagement of Indigenous Peoples and Local Communities in REDD-Plus.* Washington, DC: International Union for the Conservation of Nature.

IWG-IFR (2009). Report of the informal working group on interim finance for REDD+ (accessed 17 May 2013 from www.regjeringen.no/upload/MD/Vedlegg/Klima/klima_skogprosjektet/iwg/Report%20of%20the%20Informal%20Working%20Group%20on%20Interim%20Finance%20for%20REDD+%20_IWG%20IFR_Final.pdf).

Jackson, J. H. (1992). World trade rules and environmental policies: Congruence or conflict? *Washington and Lee Law Review*, 49(4), 1227–70.

Jacobs, R. W. (2005). Treading deep waters: Substantive law issues in Tuvalu's threat to sue the United States in the International Court of Justice. *Pacific Rim Law and Policy Journal*, 14, 103–28.

Jacobson, J. L. (1988). Environmental refugees: A yardstick of habitability. *Worldwatch Paper 86*. Washington, DC: Worldwatch Institute.

Jakobeit, C. and Methmann, C. (2007). *Klimaflüchtlinge – Die Verleugnete Katastrophe.* Hamburg: Greenpeace.

Jänicke, M., Mönch, H., Ranneberg, T. and Simonis, U. E. (1989). Economic structure and environmental impacts: East–West comparisons. *The Environmentalist*, 9(3), 171–83.

Jassanoff, S. (1990). *The Fifth Branch, Scientific Advisers as Policy Makers.* Cambridge: Harvard University Press.

Johns, T., Merry, F., Stickler, C. et al. (2008). A three-fund approach to incorporating government, public and private forest stewards into a REDD funding mechanism. *International Forestry Review*, 10(3), 458–64.

Johnson, J. (2012). Doha climate talks open amid warnings of calamity. *Irish Examiner*, 27 November 2012 (accessed 13 May 2013 from www.irishexaminer.com/world/doha-climate-talks-open-amid-warnings-of-calamity-215177.html).

Joint Statement of Academy of Sciences (2001). *The Science of Climate Change*. London: The Royal Society, 18 May 2001 (accessed 13 May 2013 from http://royalsociety.org/policy/publications/2001/science-climate-change/).

Junne, G. (1992). Beyond regime theory. *Acta Politica*, 27(1), 9–28.

Kaminskaitė-Salters, G. (2011). Climate Change Litigation in the UK: Its feasibility and prospects. In *Climate Change Liability*, ed. M. Faure and M. Peeters. Cheltenham: Edward Elgar, pp. 165–88.

Kanowski, P. J., McDermott, C. L. and Cashore, B. W. (2011). Implementing REDD+: lessons from analysis of forest governance. *Environmental Science & Policy*, 14(2), 111–7.

Kaul I., Conceicao, P., Le Goulven, K. and Mendoza, R. (eds.) (2003). *Providing Global Public Goods: Managing Globalization*. New York: Oxford University Press.

Keith, D. W. (2000). Geoengineering the climate: history and overview. *Annual Review of Energy and the Environment*, 25, 245–84.

Kelman, I. and West, J. (2009). Climate change and small island developing states: a critical review. *Ecological and Environmental Anthropology*, 5(1), 1–16.

Keohane, R. O. (1984). *After Hegemony: Cooperation and Discord in the World Political Economy*. Princeton: Princeton University Press.

Khosla, M. (2007). The TWAIL Discourse: The emergence of a new phase. *International Community Law Review*, 9, 291–304.

Kirchmeier, F. (2006). The right to development: Where do we stand? Dialogue on Globalization Occasional Papers No. 23. Geneva: Friedrich-Ebert-Stiftung.

Knox, J. H. (2009). Linking human rights and climate change at the United Nations. *Harvard Environmental Law Review*, 33, 477–98.

Kohona, P. (2002). The international rule of law and the role of the United Nations. *The International Lawyer*, 36(1), 1131–44.

Kolstad, C. (2005). Piercing the veil of uncertainty in transboundary pollution agreements. *Environmental and Resource Economics*, 31, 21–34.

Kolstad, C. and Ulph, A. (2008). Learning and international environmental agreements. *Climatic Change*, 89, 125–41.

Koskenniemi, M. (1990). The politics of international law. *European Journal of International Law*, 11(1), 4–32.

Koskenniemi, M. (2009). The politics of international law – 20 years later. *European Journal of International Law*, 20(1), 7–19.

Kosolapova, E. (2011). Liability for climate change-related damage in domestic courts: claims for compensation in the USA. In *Climate Change Liability*, ed. M. Faure and M. Peeters. Cheltenham: Edward Elgar, pp. 189–205.

Krasner, S. D. (1982). Structural causes and regime consequences: Regimes as intervening variables. *International Organization*, 36(2), 185–206.

Krause, F., Bach, W. and Koomey, J. (1989). Energy Policy in the Greenhouse. Vol. I of *International Project for Sustainable Energy Paths*. London: Earthscan Publications.

Kravchenko, S. and Bonine, J. E. (2012). Interpretation of human rights for the protection of the environment in the European Court of Human Rights. *Global Business and Development Law Journal*, 25, 245–87.

Kuntjoro-Djakti, H., Nugraha, D., Purnomo, A. and Agus, P. S. (1995). Indonesia and international climate change regime – tug of war between the state and the people; a critical response to the research report: International policies to address the greenhouse effect. In *Research Project Reviews and Commentaries of the Report: International Policies to Address the Greenhouse Effect, Working Paper 4, E1–22*. Amsterdam: Department of International Relations and Public International Law, University of Amsterdam and Institute for Environmental Studies, Vrije Universiteit.

Kuo, C., Lindberg, C. and Thomson, D. J. (1990). Coherence established between atmospheric carbon dioxide and global temperature. *Nature*, 343, 709–14.

Kwaka, E. (1994). The rule of law and global governance in the twenty-first century. *Proceedings of the 6th Annual Conference of the African Society of International and Comparative Law*, Kampala, 3–1.

Lafferty, W. M. (1996). The politics of sustainable development: Global norms for national development. *Environmental Politics*, 5(2), 185–208.

Latham, J. (1990). Control of global warming? *Nature*, 347, 339–40.

Latham, J., Rasch, P., Chen, C.-C. et al. (2008). Global temperature stabilization via controlled albedo enhancement of low-level maritime clouds. *Philosophical Transactions of the Royal Society A*, 366(1882), 3969–87.

Law, D. S. and Versteeg, M. (2011). The evolution and ideology of global constitutionalism. *California Law Review*, 99(5), 1163–257.

Lehtonen, M. (2004). The environmental-social interface of sustainable development: Capabilities, social capital, institutions. *Ecological Economics*, 49(2), 199–214.

Lenton, T. M. and Vaughan, N. E. (2009). The radiative forcing potential of different climate engineering options. *Atmospheric Chemistry and Physics*, 9, 5539–61.

Lenton, T. M., Held, H., Kriegler, E. et al. (2008). Tipping elements in the earth's climate system. *Proceedings of the National Academy of Sciences*, 105(6), 1786–93.

Li, H., Gupta, J. and van Dijk, M. P. (2013). China's drought strategies in rural areas along the Lancang River. *Water Policy*, 15(1), 1–18.

Limon, M. (2009). Human rights and climate change: Constructing a case for political action. *Harvard Environmental Law Review*, 33(2), 439–76.

Linnerooth-Bayer, J. and Mechler, R. (2006). Insurance for assisting adaptation to climate change in developing countries: a proposed strategy. *Climate Policy*, 6(6), 621–36.

Lipanovich, A. (2005). Smoke before oil: Modelling a suit against the auto and oil industry on the tobacco tort litigation is feasible. *Golden Gate University Law Review*, 35, 429–89.

Liu, J. (2011). The Role of ICAO in regulating the greenhouse gas emissions of aircraft. *Carbon and Climate Law Review*, 5(4), 417–31.

Lomborg, B. (2001). *The Sceptical Environmentalist*. Cambridge: Cambridge University Press.

Lovett, J.C., Hofman, P.S., Morsink, K. et al. (2009). Review of the 2008 UNFCCC meeting in Poznan. *Energy Policy*, 37, 3701–5.

Lyster, R. (2011). REDD+, transparency, participation and resource rights: The role of law. *Environmental Science & Policy*, 14(2), 118–26.

MA (Millennium Ecosystem Assessment) (2005). *Ecosystems and Human Well-being: Synthesis*. Washington, DC: Island Press.

Mace, M. J. (2005). Funding for adaptation to climate change: UNFCCC and GEF developments since COP-7. *Review of European Community and International Environmental Law*, 14(3), 225–46.

Malenbaüm, W. (1978). *World Demand for Raw Materials in 1985–2000*. New York: McGraw-Hill.

Malnes, R. (1995). 'Leader' and 'Entrepreneur' in international negotiations: A conceptual analysis. *European Journal of International Relations*, 1(1), 87–112.

Mank, B. C. (2005). Standing and global warming: Is injury to all, injury to none? *Environmental Law*, 35, 1–83.

Manne, A. and Richels, R. (1997). On stabilizing CO_2 concentrations – cost-effective emission reduction strategies. *Environmental Modelling and Assessment*, 2, 251–65.

Marburg, K. L. (2001). Combating the impacts of global warming: A novel legal strategy. *Colorado Journal of International Environmental Law and Policy, 2001 Yearbook*, 171–80.

Mather, A. S. (1992). The forest transition. *Area*, 24(4), 367–79.

Matsui, Y. (2002). Some aspects of the principle of 'Common but differentiated responsibilities'. *International Environmental Agreements: Politics, Law and Economics*, 2(2), 151–70.

Maxwell, S. (2012). *Chairman's Essay: Inclusive Green Growth: The Pathway to Sustainable Development. A Review of the World Bank Policy Paper by Simon Maxwell, Executive Chairman.* London: The Climate and Development Knowledge Network.

Maya, R. S. and Churie, A. (1997). Critique of the African approach to JI: Bargaining or posturing? In *Joint Implementation: Carbon Colonies or Business Opportunities?*, ed. R. S. Maya and J. Gupta. Harare: South Centre for Energy and Environment.

McDermott, C. L., Levin, K. and Cashore, B. (2011). Building the forest-climate bandwagon: REDD+ and the logic of problem amelioration. *Global Environmental Politics*, 11(3), 85–103.

McElwee, P. (2004). You say illegal, I say legal: The relationship between 'illegal' logging and land tenure, poverty, and forest use rights in Vietnam. *Journal of Sustainable Forestry*, 19 (1–3), 97–135.

McInerney-Lankford, S. (2009). Climate change and human rights: An introduction to legal issues. *Harvard Environmental Law Review*, 33, 431–7.

Meadows, D. L. (1972). *The Limits to Growth: Report to the Club of Rome.* Utrecht: Spectrum.

Mearns, R. and Norton, A. (2010). *Social Dimensions of Climate Change: Equity and Vulnerability in a Warming World.* Washington, DC: The World Bank, 319 pp.

Mebratu, D. (1998). Sustainability and sustainable development: Historical and conceptual review. *Environmental Impact Assessment Review*, 18(6), 493–520.

Meckling, J. (2011). The globalization of carbon trading: Transnational business coalitions in climate politics. *Global Environmental Politics*, 11(2), 26–50.

Meier, G. M. (2001). Ideas for development. In *Frontiers of Development Economics: The Future in Perspective*, ed. G. M. Meier and J. E. Stiglitz. Oxford/Washington, DC: Oxford University Press and World Bank, pp. 1–12.

Mendelsohn, R. and Neumann, J. E. (ed.) (1999). *The Economic Impact of Climate Change on the Economy of the United States.* Cambridge: Cambridge University Press.

Mendelsohn, R. and Nordhaus, W. (1999). The impact of global warming on agriculture. A Ricardian analysis: Reply. *American Economic Review*, 89(4), 1046–8.

Mendelsohn, R., Nordhaus, W. and Shaw, D. (1994). The impact of global warming on agriculture: A Ricardian analysis. *American Economic Review*, 84(4): 753–71.

Metz, B. (2010). *Controlling Climate Change.* Cambridge: Cambridge University Press.

Meyer, A. and Cooper, T. (1995). A recalculation of the social costs of climate change. An occasional paper in *The Ecologist* series, London.

Michaelowa, A. (2009). Will the CDM become a victim of its own success? Reform options for Copenhagen. In *Beyond Copenhagen: A Climate Policymaker's Handbook*, ed. J. Delgado and S. Gardner. Brussels: Bruegel, pp.31–40 (accessed from www.bruegel.org/publications/publication-detail/publication/323-beyond-copenhagen-a-climate-policymakers-handbook/).

Michaelowa, A., Tangen, K. and Hasselknippe, H. (2005). Issues and options for the post 2012 climate architecture: An overview. *International Environmental Agreements: Politics, Law and Economics*, 5, 5–24.

Mickelson, K. (2008). Taking stock of TWAIL histories. *International Community Law Review*, 10(4), 355–62.

Miliband, D. (2006). Building an environmental union. Speech by Rt Hon. David Miliband, MP, Secretary of State for Environment, Food and Rural Affairs, Berlin, 19 October 2006 (accessed 19 May 2013 from www.gov-news.org/gov/uk/news/39building_environmental_union39_speech_by/63280.html).

Mimura, N., Nurse, L., McLean, R. et al. (2007). Small islands. In *Climate Change 2007: Impacts, Adaptation and Vulnerability. Contribution of Working Group II to the Fourth Assessment Report of the Intergovernmental Panel on Climate Change*. Cambridge: Cambridge University Press, pp. 687–716.

Mintzer, I. M. (1994). Institutional options and operational challenges in the management of a joint implementation regime. In *Criteria for Joint Implementation under the Framework Convention on Climate Change*, ed. K. Ramakrishna. Massachusetts: Woods Hole Research Center, pp. 41–50.

Mitchell, J. F. B. and Johns, T. C. (1997). On modification of global warming by sulfate aerosols. *Journal of Climate*, 10(2), 245–67.

Mitchell, J. F. B., Karoly, D.J., Hegerl, G.C., et al. (2001). Detection of climate change and attribution of causes. In *Climate Change 2001: The Science of Climate Change. Contribution of Working Group I to the Third Assessment Report of the Intergovernmental Panel on Climate Change*, eds J. T. Houghton, D. J. Ding, M. Griggs et al. Cambridge: Cambridge University Press, pp. 694–738.

Mitra, A. P. (ed.) (1992a). *Global Change: Greenhouse Gas Emission in India, 1992 Update*. Scientific Report No. 4. New Delhi: Council of Scientific and Industrial Research.

Mitra, A. P. (1992b). Scientific basis for response of developing countries. In *Indo-British Symposium on Climate Change*, 15–17 January 1992, New Delhi, pp. 202–6.

Mol, A. P. J. and Sonnenfeld, D. A. (2000). Ecological modernisation around the world: An introduction. *Environmental Politics*, 9(1), 1–14.

Mooney, C. (2006). *The Republican War on Science*. New York: Basic Books.

Muller, A. (2007). How to make the clean development mechanism sustainable: The potential of rent extraction. *Energy Policy*, 35, 3203–12.

Müller, B. (2002). *Equity in Climate Change: The Great Divide*. Oxford: Oxford Institute for Energy Studies.

Mumma, A. and Hodas, D. (2008). Designing a global post-Kyoto climate change protocol that advances human development. *Georgetown International Environmental Law Review*, 20(4): 619–44.

Muñoz, M., Thrasher, R. and Najam, A. (2009). Measuring the negotiating burden of multilateral environmental agreements. *Global Environmental Politics*, 9(4), 1–13.

Murphy, D. M. (2009). Effect of stratospheric aerosols on direct sunlight and implications for concentrating solar power. *Environmental Science & Technology*, 43(8), 2784–6.

Mwandosya, M. J. (1999). Survival Emissions: A Perspective from the South on Global Climate Change Negotiation. Tanzania: DUP (1996) Ltd and The Centre for Energy, Environment, Science and Technology.

Myers, N. (2002). Environmental refugees: a growing phenomenon of the 21st century. *Philosophical Transactions of the Royal Society B: Biological Sciences*, 29, 357(1420): 609–13.

Myers, N. and Kent, J. (1995). *Environmental Exodus: An Emergent Crisis in the Global Arena*. Washington, DC: Climate Institute.

Nationaal Platform Rio+20 (2011). *Priorities for a Sustainable Future: Civil Society Outcomes for the Netherlands Rio+20 Preparations* (accessed 12 December 2012 from www.nprio2012.nl/wp-content/uploads/2011/06/NPRio+20-priorities-for-a-sustainable-future.pdf).

Neumann, F. L. (1996). The change in the function of law in modern society. In *The Rule of Law Under Siege*, ed. W. E. Scheuerman. Berkeley: University of California Press.

Newman, E., Tirman, J. and Thakur, R. (ed.) (2006). *Multilateralism under Challenge? Power, Normative Structure and World Order*. Tokyo: United Nations University Press.

Nikitina, E. (2001). Russia: Climate Policy Formulation and Implementation During the 1990s. *Climate Policy*, 1, 289–308.

Noordwijk Declaration (1989). Noordwijk Declaration on Climate Change. In *Noordwijk Conference Report: Vol. I*, ed. P. Vellinga, P. Kendall and J. Gupta. The Hague: Netherlands Ministry of Housing, Physical Planning and Environment.

Nordhaus, W. D. (2008). *A Question of Balance: Weighing the Options on Global Warming Policies*. New Haven: Yale University Press.

NRC (2010). *Advancing the Science of Climate Change. America's Climate Choices: Panel on Advancing the Science of Climate Change*. Washington, DC: The National Academies Press.

Nurkse, R. (1953). *Problems of Capital Formation in Developing Countries*. Oxford: Blackwell.

Oberthür, S. and Ott, H. E. (1999). *The Kyoto Protocol: International Climate Policy for the 21st Century*. Berlin: Springer Verlag.

Okafor, O. C. (2005). Newness, imperialism and international legal reform in our time: A TWAIL perspective. *Osgoode Hall Law Journal*, 43(1/2), 171–91.

Okafor, O. C. (2008). Critical Third World Approaches to International Law (TWAIL): Theory, methodology or both? *International Community Law Review*, 10(4), 371–8.

Okubo, Y and Michaelowa, A. (2010). Effectiveness of subsidies for the Clean Development Mechanism: Past experiences with capacity building in Africa and LDCs. *Climate and Development*, 2(1), 30–49.

Olsthoorn, X. and Wieczorek, A. (ed.) (2006). *Understanding Industrial Transformation: Views from Different Disciplines*. Dordrecht: Kluwer Academic Publishers, pp. 53–74.

O'Neill, B. C. and Oppenheimer, M. (2002). Dangerous climate impacts and the Kyoto Protocol. *Science*, 296(5575), 1971–2.

Oppenheimer, M. and Petsonk, A. (2003). Global warming: The intersection of long-term goals and near-term policy. In *Climate Policy for the 21st Century: Meeting the Long-Term Challenge of Global Warming*, ed. D. Michel. Washington, DC: Center for Transatlantic Relations, Johns Hopkins University, pp. 79–112.

Oppenheimer, M. and Petsonk, A. (2005). Article 2 of the UN Framework Convention on Climate Change (UNFCCC): Historical origins, recent interpretations. *Climatic Change*, 73(3), 195–226.

Opschoor, J. B. (2009). Sustainable development and a dwindling carbon space. Public Lecture Series 2009, No. 1. The Hague: Institute of Social Studies.

Opschoor, J. B. (2009). Sustainability. In *Handbook on Economics and Ethics*, ed. I. van Staveren and J. Peil. Cheltenham: Edward Elgar, pp. 531–8.

Osafsky, H. M. (2011–2012). Litigation's role in the path of the U.S. federal climate change regulation: Implications of AEP v. Connecticut. *Valparaiso University Law Review*, 46, 447–657.

Ostrom, E. (2011). Reflections on 'Some unsettled problems of irrigation'. *American Economic Review*, 101, 49–63.

Ott, K., Klepper, G., Lingner, S. et al. (2004). *Reasoning Goals of Climate Protection. Specification of Art. 2 UNFCCC*. Berlin: Umweltbundesamt.

Parikh, J. (1992). IPCC strategies unfair to the South. *Nature*, 360, 507–8.

Parker, C., Mitchell, A., Trivedi, M. et al. (2009). *The Little REDD+ Book: An Updated Guide to Governmental and Non-governmental Proposals for Reducing Emissions from Deforestation and Degradation*. Oxford: Global Canopy Programme.

Pedroni, L., Dutschke, M., Streck, C. and Porrúa, M. E. (2009). Creating incentives for avoiding further deforestation: the nested approach. *Climate Policy*, 9(2), 207–20.

Peeters, M. (2011). The regulatory approach of the EU in view of liability for climate change damage. In *Climate Change Liability*, ed. M. Faure and M. Peeters. Cheltenham: Edward Elgar, pp. 90–133.

Penalver, E. M. (1998). Acts of God or toxic torts? Applying tort principles to the problem of climate change. *Natural Resources Journal*, 38, 563–9.

Peters, G. P., Marland, G., Le Quéré, C. et al. (2012). Rapid growth in CO_2 emissions after the 2008–2009 global financial crisis. *Nature Climate Change*, 2, 2–4.

Petit, J. R., Jouzel, J., Raynaud, D. et al. (1999). Climate and atmospheric history of the past 420,000 years from the Vostok ice core, Antarctica. *Nature*, 399, 429–36.

Phelps, J., Webb, E. L. and Agrawal, A. (2010). Does REDD+ threaten to recentralize forest governance? *Science*, 328(5976), 312–13.

Pierro, R. and Desai, B. (2008). Climate insurance for the poor: Challenges for targeting and participation. *IDS Bulletin*, 39(4), 123–9.

Prasad, M. (1990). Speech. In *Noordwijk Conference Report*, Vols I and II, ed. P. Vellinga, P. Kendall, J. Gupta et al. The Netherlands: Ministry of Housing, Physical Planning and Environment.

Prescott, J. (2000). It's time for a deal. *International Herald Tribune*, November 18–19.

Priem, H. N. A. (1995). De CO_2 ideologie, wetenschap en onderwijs. *NRC Handelsblad*, 6 July, 1–2.

Rahmstorf, S. (2009). Anthropogenic climate change: Revisiting the facts. In E. Zedillo (ed.), *Global Warming: Looking Beyond Kyoto*. Washington, DC: Brookings Institution Press, pp. 34–53.

Rajagopal, B. (2006). Counter-hegemonic international law: Rethinking human rights and development as a Third World strategy. *Third World Quarterly*, 27(5), 767–83.

Rajamani, L. (2000). The principle of common but differentiated responsibility and the balance of commitments under the climate regime. *Review of European Community and International Environmental Law*, 9(2), 120–31.

(2006). Differential Treatment in International Environmental Law. Oxford University Press.

Rasch, P. J., Crutzen, P. J. and Coleman, D. B. (2008). Exploring the geoengineering of climate using stratospheric sulfate aerosols: the role of particle size. *Geophysical Research Letters*, 35, L02809.

Rauniyar, G. and Kanbur, R. (2010a). Inclusive growth and inclusive development: A review and synthesis of Asian Development Bank literature. *Journal of the Asia Pacific Economy*, 15(4), 455–69.

Rauniyar, G. and Kanbur, R. (2010b). Inclusive development: Two papers on conceptualization, application, and the ADB perspective (accessed 12 December 2012 from http://kanbur.dyson.cornell.edu/papers/ADBCompendiumInclusiveDevelopment. pdf).

Raworth, J. (2012). A safe and just space for humanity: Can we live within the doughnut? Oxfam Discussion Papers (accessed 13 May 2013 from www.oxfam.org/sites/www. oxfam.org/files/dp-a-safe-and-just-space-for-humanity-130212-en.pdf).

Read, P. (2008). Biosphere carbon stock management: addressing the threat of abrupt climate change in the next few decades: an editorial essay. *Climatic Change*, 87, 305–20.

Reed, E., Simon, B. and Jung-a, S. (2009). Trade-in incentives not long-term solution. *Financial Times*, 27 March 2009.

Revesz, R. and Livermore, M. (2008). *Retaking Rationality: How Cost Benefit Analysis Can Better Protect the Environment and Our Health*. New York: Oxford University Press.

Reynolds, N. B. (1989). Grounding the rule of law. *En Ratio Iuris*, 2, 1–16.

Rezessy, S., Dimitrov, K., Urge-Vortsatz, D. et al. (2006). Municipalities and energy efficiency in countries in transition: Review of factors that determine municipal involvement in the markets for energy services and energy efficient equipment, or how to augment the role of municipalities as market players. *Energy Policy*, 34, 223–37.

Ricke, K. L., Morgan, M. G. and Allen, M. R. (2010). Regional climate response to solar-radiation management. *Nature Geoscience*, 3, 537–41.

Robinson, J. (2004). Squaring the circle? Some thoughts on the idea of sustainable development. *Ecological Economics*, 48, 369–84.

Rockström, J., Steffen, W., Noone, K. et al. (2009). Planetary boundaries: Exploring the safe operating space for humanity. *Ecology and Society*, 14(2), 32–65.

Roger, C. and S. Belliethathan (2013). Africa in the Global Climate Change Negotiations, 1992–2012. CIGI Africa Initiative Discussion Paper.

Roger, C., Alger, J. and Allan, J. (2013). UNFCCC Participation and Submissions Dataset Version 1.0, available with authors.

Rosenstein-Rodan, P. (1943). Problems of industrialization in Eastern and South-Eastern Europe. *The Economic Journal*, 53(210/211), 202–11.

Rostow, W. W. (1956). The take-off into self-sustained growth. *The Economic Journal*, 66(261), 25–48.

Rostow, W. W. (1960). *The Stages of Economic Growth: A Non-communist Manifesto*. Cambridge: Cambridge University Press.

Rotmans, J., Kemp, R., van Asselt, M. et al. (2000). *Transities en Transitiemanagement: De Casus van een Emissiearme Energievoorziening*. Maastricht: ICIS, Merit.

Rudel, T. K. Coomes, O. T., Moran, E. et al. (2005). Forest transitions: towards a global understanding of land use change. *Global Environmental Change*, 15, 23–31.

Sachs, I. (2004). Inclusive development strategy in an era of globalization. Working Paper No. 35. Geneva: International Labour Office, Policy Integration Department, World Commission on the Social Dimension of Globalization.

Sachs, W. (1999). Sustainable development and the crisis of nature: On the political anatomy of an oxymoron. In *Living with Nature*, ed. F. Fischer and M. Hajer. Oxford: Oxford University Press, pp. 23–42.

Salomon, M. E. (2009). Poverty, privilege and international law: The Millennium Development Goals and the guise of humanitarianism. *German Yearbook of International Law*, 51, 39–73.

Salter, S., Sortino, G. and Latham, J. (2008). Sea-going hardware for the cloud albedo method of reversing global warming. *Philosophical Transactions of the Royal Society A*, 366(1882), 3989–4006.

Sand, P. (1990). *Lessons Learned in Global Environmental Governance*. New York: World Resources Institute.

Sandelius, W. (1933). National sovereignty versus the rule of law. *American Political Science Review*, 25(1), 1–20.

Sands, P. (1992). The United Nations Framework Convention on Climate Change. *Review of European Community and International Environmental Law*, 1(3), 270–7.

Sands, P. (2003). *Principles of International Environmental Law*. Cambridge: Cambridge University Press.

Santarius, T., Arens, C., Eichhorst, U. et al. (2009). Pit stop Poznan. An analysis of negotiations on the Bali Action Plan at the stopover to Copenhagen. *Journal of European Environmental Planning Law*, 6(1), 75–96.

Santiso, C. (2001). Good governance and aid effectiveness: The World Bank and conditionality. *The Georgetown Public Policy Review*, 7(1), 1–22.

Schelling, T. C. (1997). The cost of combating global warming: Facing the trade-offs. *Foreign Affairs*, 76(6), 8–14.

Schneider, L. (2011). Perverse incentives under the CDM: An evaluation of HFC 23 destruction projects. *Climate Policy*, 11, 851–64.

Schneider, S. H. (1996). Geoengineering: Could – or should – we do it? *Climatic Change*, 33(3), 291–302.

Schrijver, N. (1995). Sovereignty over natural resources: Balancing rights and duties in an interdependent world. PhD Thesis at Rijksuniversiteit Groningen, Netherlands.

Schwarzkopf, M. D. and Ramaswamy, V. (2008). Evolution of stratospheric temperature in the 20th century. *Geophysical Research Letters*, 35, L03705.

Scruggs, L. and Benegal, S. (2012). Declining public concern about climate change: Can we blame the recession? *Global Environmental Change*, 22, 505–16.

Sebenius, J. K. (1993). The Law of the Sea Conference: Lessons for negotiations to control global warming. In *International Environment Negotiations*, ed. G. Sjostedt. Newbury Park/London/New Delhi: Sage, pp. 189–216.

Seif El-Dawla, A. (2002). Good governance after September 11th 2001: A new dimension for international development cooperation. *Griffin's View*, 3(1), 15–26.

Selden, T. M. and Song, D. (1996). Environmental quality and development: Is there a Kuznets curve for air pollution emissions? *Journal of Environmental Economics and Management*, 27(2), 147–62.

Sen, A. (1999). *Development as Freedom*. Oxford: Oxford University Press.

Senona, J. M. (2009). EPAs and the Doha Round: Development or discontent. *Journal of International Trade Law and Policy*, 8(1), 60–83.

Shackley, S. and Skodvin, T. (1995). IPCC gazing and the interpretive social sciences. *Global Environmental Change*, 5(3), 175–80.

Sharma, V. (2001). Utilitarian rationalism on thin ice. *The Economic Times*, 6 January, India.

Shiva, V. and Bandyopadhyay, J. (1986). Environmental conflicts and public interest science. *Economic and Political Weekly*, XXI(2), 84–90.

Shum, R. Y. (2012). Effects of economic recession and local weather on climate change attitudes. *Climate Policy*, 12(1), 38–49.

Simma, B. (1994). From bilateralism to community interest in international law. *Recueil des Cours*, 217–50.

Sinden, A. (2005). In defense of absolutes: Combating the politics of power in environmental law. *Iowa Law Review*, 90, 1405–512.

Sinden, A. (2007). Climate change and human rights. *Journal of Land, Resources and Environmental Law*, 27, 255–72.

Slade, T. (2003). The making of international law. The role of Small Island States. *Temple International & Comparative Law Journal*, 17, 531–43.

Slade, T. N. (1998). National perspectives – ratification and implementation strategies: The view from the Small Island States. Paper presented at the conference: *Climate after Kyoto – Implications for Energy*, Chatham House, 5–6 February 1988.

Smit, B., Pilifosova, O. (2001). Adaptation to climate change in the context of sustainable development and equity. In *Climate Change 2001: Impacts, Adaptation, and Vulnerability – Contribution of Working Group II to the Third Assessment Report of the Intergovernmental Panel on Climate Change*, ed. J. J. McCarthy, O. F. Canzianni, N. A. Leary et al. Cambridge: Cambridge University Press, pp. 876–912.

Smit, B., Pilifosova, O. Burton, I. et al. (2001). Adaptation to climate change in the context of sustainable development and equity. In *Climate Change 2001: Impacts, Adaptation, and Vulnerability – Contribution of Working Group II to the Third Assessment Report of the Intergovernmental Panel on Climate Change*. Cambridge: Cambridge University Press.

Smith, J. and Shearman, D. (2006). *Climate Change Litigation: Analysing the Law, Scientific Evidence and Impact on the Environment, Health and Property*. Adelaide: Presidian Legal Publications.

Smith, J. B., Dickinson, T., Donahue, J. D. B. et al. (2011). Development and climate change adaptation funding: Coordination and integration. *Climate Policy*, 11, 987–1000.

Smith, K. (1995). The natural debt: North and South. In *Manuscript for Climate Change: Developing Southern Hemisphere Perspectives*, ed. T. A. Giambelluca and A. Henderson-Sellers (1996). Sussex: J. Wiley and Sons.

Smith, L. C. (2010). *The World in 2050: Four Forces Shaping Civilization's Northern Future*. Dutton: Penguin Group.

Sornarajah, M. (2001). The Asian perspective to international law in the age of globalization. *Singapore Journal of International & Comparative Law*, 5(2), 284–313.

Stavins, R. N. (2011). The Problem of the Commons: Still unsettled after 100 years. *American Economic Review*, 101(1), 81–108.

Sterling, S. (2010–2011). Transformative learning and sustainability: Sketching the conceptual ground. *Learning and Teaching in Higher Education*, 5, 17–33.

Stern, N. (2006). *The Economics of Climate Change: The Stern Review*. Cambridge: Cambridge University Press.

Stone, C. D. (2004). Common but differentiated responsibilities in international law. *American Journal of International Law*, 98(2), 276–301.

Strassburg, B., Kerry Turner, R., Fisher, B. et al. (2009). Reducing emissions from deforestation: The 'combined incentives' mechanism and empirical simulations. *Global Environmental Change*, 19, 265–78.

Swart, R. (1994). Climate change: Managing the risks. PhD thesis at Vrije Universiteit, Amsterdam.

Sydney Centre for International and Global Law (2004). Global climate change and the Great Barrier Reef: Australia's obligations under the World Heritage Convention. A report prepared by the Sydney Centre for International and Global Law, Faculty of Law, University of Sydney, Australia (accessed 14 May 2013 from http://sydney.edu.au/law/scigl/SCIGLFinalReport21_09_04.pdf).

Tamanaha, B. (2004). *On the Rule of Law: History, Politics and Theory*. Cambridge: Cambridge University Press.

Tearfund (2006). Overcoming the barriers: Mainstreaming climate change adaptation in developing countries. Tearfund Climate Change Briefing Paper 1. Teddington: Tearfund.

Teller, E., Hyde, R. and Wood, L. (2002). *Active Climate Stabilization: Practical Physics-Based Approaches to Prevention of Climate Change*. Preprint UCRL-JC-148012. Livermore: Lawrence Livermore National Laboratory.

Thackeray, R. W. (2004). Struggling for air: The Kyoto Protocol, citizens' suits under the Clean Air Act, and the United States options for addressing global climate change. *Indiana International and Comparative Law Review*, 14, 855–903.

Thatcher, M. (1988). Speech to the Royal Society, 27 September 1988 (accessed 19 May 2013 from www.margaretthatcher.org/document/107346).

The US Guardian (2010). The US Embassy Cables (accessed from www.guardian.co.uk/environment/2010/dec/03/wikileaks-us-manipulated-climate-accord).

Thompson, W. S. (1929). Population. *American Journal of Sociology*, 34, 959–75.

Thorbecke, E. (2006). The evolution of the development doctrine: 1950–2005. UNU-WIDER Discussion Paper 2006/155. Helsinki: World Institute for Development Economics Research of the United Nations University.

Tolba, M. K. (1989). Our biological heritage under siege. *Bioscience*, 39, 725–8.

Toth, F. L. (1995). Practice and progress in integrated assessment of climate change: A workshop overview. *Energy Policy*, 23(4/5), 253–67.

Trail Smelter Case (1941). US versus Canada, Reports of Arbitral Awards, Vol. III, pp. 1905–82 (accessed from http://untreaty.un.org/cod/riaa/cases/vol_iii/1905-1982.pdf).

Tsagourias, N. (2003). Globalization, order and the rule of law. *Finnish Yearbook of International Law*, IX, 247–64.

Ulden, A. P. van (1995). Analyse van de Stellingen van Böttcher. Paper presented at the *Workshop over Klimaatverandering*, 21 March 1995. Amsterdam: World Information Service on Energy.

Underdal, A. (1994). Leadership theory: rediscovering the arts of management. In *International Multilateral Negotiating: Approaches to the Management of Complexity*, ed. W. I. Zartman. San Francisco: IIASA/Jossey-Bass, pp.178–97.

UNDP (2007). *Fighting climate change: Human solidarity in a divided world. UNDP Human Development Report 2007–2008*. New York: Palgrave Macmillan.

UNDP (2011). *Human Development Report, Sustainability and Equity: A Better Future for All*. Basingstoke: Palgrave Macmillan.

UNEP (2010). *Towards a Green Economy: Pathways to Sustainable Development and Poverty Eradication* (accessed from www.unep.org/greeneconomy/Portals/88/documents/ger/GER_synthesis_en.pdf).

UNEP (2012). Emissions Gap Report, UNEP.

UNGA (2007). Declaration on the Rights of Indigenous Peoples, A/RES/66/142.

UNIDO (2008). *Public Goods for Economic Development*. Vienna: UNIDO.

Van den Bergh, J. C. (2010). Safe climate policy is affordable – 12 reasons. *Climatic Change*, 101, 339–85.

van der Grijp, N., Bergsma, E. and Gupta, J. (2012). The Dutch focus: A Delta Act for Climate Legislation. In *Climate Law in EU Member States: Towards National Legislation for Climate Protection*, ed. M. Peeters, M. Stallworthy and J. de C. de Larragan. Cheltenham/Northampton: Edward Elgar, pp. 312–30.

van der Sluijs, J. P. (1997). Anchoring amid uncertainty: On the management of uncertainties in risk assessment of anthropogenic climate change. PhD thesis at the University of Utrecht.

van der Tak, J., Haub, C. and Murphy, E. (1979). Our population predicament: A new look. *Population Bulletin*, 34(5), 3–48.

van Dijk, C. (2011). Civil liability for global warming in the Netherlands. In *Climate Change Liability*, ed. M. Faure and M. Peeters. Cheltenham: Edward Elgar, pp. 206–26.

van Dijk, P. (1987). Normative force and effectiveness of international norms. *German Yearbook of International Law*, 30, 9–35.

Vellinga, P. (2012). *On Climate Change: Facts, Myths and Serious Questions*. Amsterdam: Uitgeverij Balans.

Vellinga, P., Kendall, P. and Gupta, J. (ed.) (1990). *Noordwijk Conference Report, Vol. II, Ministry of Housing, Physical Planning and the Environment.* The Hague: VROM.

Vellinga, P., Mills, E., Berz, G. et al. (2001). Insurance and other financial services. In *Climate Change 2001: Impacts, Vulnerability and Adaptation. Contribution of Working Group II to the Third Assessment Report of the Intergovernmental Panel on Climate Change.* Geneva: IPCC.

Verheyen, R. (2003). Climate change damage in international law. Dissertation zur Erlangung des Dr. iur Universität Hamburg, Fachbereich Rechtswissenschaft.

Virgoe, J. (2008). International governance of a possible geoengineering intervention to combat climate change. *Climatic Change,* 95, 103–13.

von Weizsäcker, E., Lovins, A. B. and Lovins, L. H. (1997). *Factor Four. Doubling Wealth – Halving Resource Use.* London: Earthscan.

Wagner, L. (1999). Negotiations in the UN Commission on Sustainable Development, International Negotiation. *International Negotiation: Journal of Theory and Practice,* 4(2), 107–31.

Waldron, J. (2011). Are sovereigns entitled to the benefit of the international rule of law? *The European Journal of International Law,* 22(2), 315–43.

Walsh, S., Tian, H., Whalley, J. et al. (2011). China and India's participation in global climate negotiations. *International Environmental Agreements,* 11, 261–73.

Warner, K. and Zakieldeen, S. A. (2012). *Loss and Damage Due to Climate Change: An Overview of the UNFCCC Negotiations. European Capacity Building Initiative* (accessed 13 May 2013 from www.climate-insurance.org/dbfs.php?file=dbfs:/WarnerZakieldeen2011_ECBI_Loss and Damage.pdf)

Warner, K., Bouwer, L. M. and Ammann, W. (2007). Financial services and disaster risk finance: examples from the community level. *Environmental Hazards,* 7(1), 32–9.

Warner, K., Ranger, N., Surminski, S. et al. (2009). *Adaptation to Climate Change: Linking Disaster Risk Reduction and Insurance.* Geneva: United Nations International Strategy for Disaster Reduction.

Watts, A. (1993). The International Rule of Law. *German Yearbook of International Law,* 36, 15–45.

WCC (World Climate Conference) (1979). *Proceedings of the World Climate Conference: A Conference of Experts on Climate and Mankind.* Geneva: World Meteorological Organization.

WCED (World Commission on Environment and Development) (1987). *Our Common Future: Brundtland Report of the World Commission on Environment and Development.* Oxford: Oxford University Press.

Weiss, C. (2006). Precaution: The willingness to accept costs to avert uncertain danger. In *Coping with Uncertainty – Modelling and Policy Issues, Lecture Notes in Economics and Mathematical Systems,* ed. K. Marti, Y. Ermoliev, M. Makowski and G. Pflug. Berlin/Heidelberg: Springer, pp. 315–30.

Weisslitz, M. (2002). Rethinking the equitable principle of common but differentiated responsibility: Differential versus absolute norms of compliance and contribution in the global climate change context. *Colorado Journal of International Environmental Law and Policy,* 13, 473–509.

WEO (2012). *World Energy Outlook 2012.* Paris: IEA.

Wettestad, J. (2000). The complicated development of EU climate policy. In *Climate Change and European Leadership: A Sustainable Role for Europe,* ed. J. Gupta and M. Grubb. Environment and Policy Series. Dordrecht: Kluwer Academic Publishers, pp. 25–46.

White House (2001). *President Bush Discusses Climate Change* (accessed 8 July 2012 from http://georgewbush-whitehouse.archives.gov/news/releases/2001/06/20010611–2. html).

White House (2004). US Climate Change Policy: The Bush Administration's actions on global climate change. Fact sheet released by the White House, Office of the Press Secretary, 19 November 2004, Washington, DC (accessed 13 May 2013 from www. state.gov/g/oes/rls/fs/2004/38641).

Wiener, J. (2009). Property and prices to protect the planet. *Duke Journal of Comparative & International Law*, 19, 515–34.

Wigley, T. M. L. (2006). A combined mitigation/geoengineering approach to climate stabilization. *Science*, 314(5798), 452–4.

Wigley, T. M. L., Richels, R. and Edmonds, J. A. (1996). Economic and environmental choices in the stabilization of atmospheric CO_2 concentrations. *Nature*, 379, 240–3.

Wingenter, O. W., Elliot, S. M. and Blake, D. R. (2007). New directions: Enhancing the natural sulphur cycle to slow global warming. *Atmospheric Environment*, 41(34), 7373–5.

Wolters, G., Swager, J. and Gupta, J. (1991). Climate change: A brief history of global, regional and national policy measures. Paper presented at the *International Global Warming Symposium Organized by the Japan Society for Air Pollution*, Tokyo, 15 November 1991.

Wong, P. P. (2011). Small Island Developing States. *WIREs Climate Change*, 2, 1–6.

Woods, N. (1999). Good governance in international organizations. *Global Governance*, 5, 39–62.

World Bank (2012). *Inclusive Green Growth. The Pathway to Sustainable Development*. Washington, DC: Worldbank (accessed 12 December 2012 from http://siteresources. worldbank.org/EXTSDNET/Resources/Inclusive_Green_Growth_May_2012. pdf).

Xie, X. and Economides, M. (2009). Great leap forward for China's wind energy. *Energy Tribune*, 30 July.

Yamin, F. (1998). The Kyoto Protocol: Origins, assessment and future challenges. *Review of European Community and International Environmental Law*, 7(2), 113–27.

Yohe, G., Neumann, J., Marshall, P. et al. (1996). The economic costs of sea-level rise on developed property in the United States. *Climate Change*, 32, 387–410.

Young, O. R. (1991). Political leadership and regime formation: On the development of institutions in international society. *International Organization*, 45(3), 201–308.

Zaelke, D. and Cameron, J. (1990). Global warming and climate change: An overview of the international legal process. *American University International Law Review*, 5(2), 249–90.

Zafrilla, J. E., López, L. A., Cadarso, M. Á. et al. (2012). Fulfilling the Kyoto protocol in Spain: A matter of economic crisis or environmental policies? *Energy Policy*, 51, 708–19.

Zhang, Z. X. (1999). Is China taking action to limit its greenhouse gas emissions? Past evidence and future prospects. In *Promoting Development While Limiting Greenhouse Gas Emissions: Trends and Baselines*, ed. J. Goldenberg and W. Reid. New York: UNDP and WRI.

Statements, declarations, treaties and UN documents

Agenda 21 (1992). Agenda 21: Action Plan for the next century. In *United Nations Conference on Environment and Development*, 3–14 June, 1992. Rio de Janeiro: UNCED.

AOSIS (1991). UN Doc. A/AC.237/Misc.1/Add.3 at 30.

AOSIS (2008). Proposal to the AWG-LCA: Multi-Window Mechanism to Address Loss and Damage from Climate Change Impacts (Paper No. 3c). In: FCCC/AWG-LCA/2008/MISC.5/Add.2 (Part I).

AOSIS Protocol (1994). AOSIS Protocol, Draft Protocol Text. A/AC./237/L.23, 27 September 1994.

Bergen Declaration (1990). UN/ECE Bergen Ministerial Declaration on Sustainable Development in the ECE Region. UN Doc. A/CONF.151/PC/10, Annex I. Bergen.

Byrd–Hagel Resolution (1997). US Senate Resolution 98 on Global Warming. Congressional Record, Senate, page S10308-S10311. 3 October 1997.

CBD (1992). United Nations Convention on Biological Diversity, adopted 5 June 1992; entry into force 29 December 1993. 31 ILM (1992), 818.

Commission on Climate Change and Development (2009). *Closing the Gaps: Report of the Commission on Climate Change and Development.* Stockholm: Swedish Ministry of Foreign Affairs.

Council of Environmental Ministers (1997). *European Community Conclusions on Climate Change,* Council of Ministers, European Union, 3 March 1997.

Declaration of Brasilia (1989). Declaration of Brasilia. Issued at the *Sixth Ministerial Meeting on the Environment in Latin America and the Caribbean,* Brasilia, 30–31 March 1989.

EC (1998). Climate change: Towards an EU post-Kyoto strategy. Communication from the Commission to the Council and the European Parliament. COM (98) 353 final, 3 June 1998.

EC (2003). Communication from the Commission to the Council, the European Parliament and the European Economic and Social Committee: Governance and Development. Brussels, 20 October 2003, COM(2003)615 final.

EC (2011). A Roadmap for Moving to a Competitive Low Carbon Economy in 2050. Commission Communication. COM(2011)112. 8 March 2011.

EC (2012). *Global Europe 2050.* Brussels: Directorate for Research and Innovation.

EU Council Conclusions (1997). Amsterdam European Council, 16/17 June 1997.

European Community (1990). Conclusions on Climate Change, October 1990. European Council.

European Environment Agency (2004). Impacts of Europe's Changing Climate. Report 2/2004, Copenhagen.

Government of India (2008). National Action Plan on Climate Change. New Delhi: Government of India.

Group of 77 South Summit (2000). *Declaration of the South Summit,* Havana, 10–14 April 2000.

Hague Declaration (1989). Declaration of the Hague. Meeting of Heads of State, 11 March 1989.

IMO (1972). *Convention on the Prevention of Marine Pollution by Dumping of Wastes and Other Matter.* Adopted on 13 November, 1972; entry into force 30 August 1975. London: London Convention.

INC4 (1991). UN Doc.A/AC.237/15, 28.

Langkawi Declaration (1989). Langkawi Declaration on the Environment. Issued by Commonwealth Heads of Government at Langkawi, Malaysia. 21 October 1989.

Malé Declaration (2007). Malé Declaration on the Human Dimension of Global Climate Change (2007). Adopted by the representatives of the Small Island Developing States, 14 November 2007 (accessed 13 May 2013 from http://www.ciel.org/Publications/Male_Declaration_Nov07.pdf).

Malé Declaration (1989). Malé Declaration on Global Warming and Sea Level Rise. Adopted by the Small States Conference on Sea Level Rise, hosted by the Government of the Maldives, 14–18 November 1989.

NAM (1989). Statement on Environment made at the Ninth Conference of Heads of State or Government of Non-Aligned Countries. NAC 9/EC/Doc. 8/Rev.3, 7 September 1989.

Nigerian Constitution (1999). CAP C23 Constitution of the Federal Republic of Nigeria (Promulgation Act) 1999.

OHCHR (2009). Report of the Office of the United Nations High Commissioner for Human Rights on the relationship between climate change and human rights. A/HRC/10/61, 15 January, 2009 (accessed 13 May 2013 from www.refworld.org/docid/498811532.html).

People's Agreement (2010). World People's Conference on Climate Change, and the Rights of Mother Earth, 22 April, Cochabamba, Bolivia.

Rio Declaration (1992). Rio Declaration on Environment and Development. Report on the United Nations Conference on Environment and Development, 3–14 June 1992, UN doc. A/CONF.151/26/Rev.1 (Vols I–III). Rio de Janeiro: UNCED.

Rio Forest Principles (1992). Non-Legally Binding Authoritative Statement of Principles for a Global Consensus on the Management, Conservation and Sustainable Development of All Types of Forests. 13 June 1992, 31 ILM 881.

Stockholm Declaration (1972). Declaration of the United Nations Conference on the Human Environment. Stockholm, 16 June 1972.

SWCC (1990). *Scientific Declaration of the Second World Climate Conference.* Geneva: World Meteorological Organization.

SWCC Ministerial Declaration (1990). *Ministerial Declaration of the Second World Climate Conference.* Geneva: World Meteorological Organization.

Tata Conference Statement (1989). *International Conference on Global Warming and Climate Change: Perspectives from Developing Countries*, 21–23 February 1989, New Delhi.

Tokyo Conference (1989). *Tokyo Conference on the Global Environment and Human Response: Toward Sustainable Development*, 11–13 September 1989.

Toronto Declaration (1988). *Conference Statement of the Conference: The Changing Atmosphere: Implications for Global Security.* Organized by the Government of Canada, 27–30 June, Toronto.

Toronto Statement (1989). *Protection of the Atmosphere: Statement of the Meeting of Legal and Policy Experts.* 22 February, 1989. Ottawa.

UN (1945). *Charter of the United Nations.* 26 June 1945, in force 24 October 1945 and amended 17 December 1963, 20 December 1965 and 20 December 1971. ICJ Acts and Documents No. 4, San Francisco.

UN (1976). *United Nations Convention on the Prohibition of Military or any Other Hostile Use of Environmental Modification Techniques.* Adopted by Resolution 31/72 of the United Nations General Assembly on 10 December 1976, entry into force 5 October 1978. Geneva.

UN (1997). United Nations Convention on the Law of the Non-Navigational Uses of International Watercourses. 36 ILM 700. 21 May 1997; not yet in force.

UN (2002). Decision 4/CP.7, Development and Transfer of Technologies (Decisions 4/CP.4 and 9/CP.5). UN Doc. FCCC/CP/2001/13/Add.1. 21 January, 2002.

UN (2008a). Decision 1/CP.13, Bali Action Plan. UN Doc. FCCC/CP/2007/6/Add.1 (14 March 2008).

UN (2008b). Reducing emissions from deforestation in developing countries: approaches to stimulate action. Draft conclusions proposed by the Chair. UN Doc. UNFCCC/SBSTA/2008/L.23. Poznan, 1–10 December 2008.

UN (2012). *Rio +20 Declaration on the Future We Want.* Resolution adopted by the General Assembly, sixty-sixth session. A/RES/66/288.

UN HRC (2010). Resolution adopted by the Human Rights Council* 15/9, Human rights and access to safe drinking water and sanitation.

UN Secretary General (2010). Progress to Date and Remaining Gaps in the Implementation of the Outcomes of the Major Summits in the Area of Sustainable Development, as well as an Analysis of the Themes of the Conference. Report of the Secretary General. A/CONF.216/PC/2. 1 April 2010.

UN Watercourses Convention (1997). *Convention on the Law of the Non-Navigational Uses of International Watercourses*, New York, 21 May 1997, Doc. A/51/869 adopted in Resolution A/RES/51/229.

UNCCD (1994). Convention to Combat Desertification. Adopted 17 June 1994, entry into force 26 December, 1996. A/AC.241/27.

UNCLOS (1982). *United Nations Convention on the Law of the Sea*. Adopted 10 December, 1982, entry into force 16 November 1994 (Article 308). Montego Bay, Jamaica.

UNDP (1997). *Governance for Sustainable Human Development* (accessed 13 May 2013 from http://mirror.undp.org/magnet/policy/).

UNFCCC (1991a). Preparation of a Framework Convention on Climate Change: Set of informal papers provided by delegations, related to the preparation of a Framework Convention on Climate Change, UN doc. A/AC.237/Misc.1/Add.3 (18 June 1991).

UNFCCC (1991b). Report of the Intergovernmental Negotiating Committee for a Framework Convention on Climate Change on the work of its fourth session, A/AC.237/15, Geneva (9–20 December 1991).

UNFCCC (1992). United Nations Framework Convention on Climate Change. Adopted on 9 May 1992, New York; entry into force 24 March 1994. 31 I.L.M. 849.

UNFCCC (1997). Report of the Ad Hoc Group on the Berlin Mandate on the work of its Sixth Session, Bonn, 3–7 March 1997. Addendum, Proposals for a protocol or another legal instrument. FCCC/AGBM/1997/3/Add.1.

UNFCCC Compilation (2011). Compilation and synthesis of fifth national communications, FCCC/SBI/2011/INF.1, 2011.

UNFCCC Secretariat (1998). Review of the Implementation of the Commitments and of other Provisions of the Convention. Note by the Secretariat. FCCC/CP/1998/11.

UNFCCC Secretariat (2008). Investment and Financial Flows to Address Climate Change: An Update. Technical Paper. UN Doc. FCCC/TP/2008/7, 26 November 2008 (accessed 13 May 2013 from http://unfccc.int/resource/docs/2008/tp/07.pdf).

UNGA (1966a). International Covenant on Civil and Political Rights. 16 December 1966, United Nations, Treaty Series, 999.

UNGA (1966b). International Covenant on Economic, Social and Cultural Rights. 16 December 1966, United Nations, Treaty Series, 993.

UNGA (1979). Convention on the Elimination of All Forms of Discrimination Against Women. 18 December 1979, United Nations, Treaty Series, 1249.

UNGA (1986). Declaration on the Right to Development. United Nations General Assembly Resolution A/RES/41/128, 4 December, 1986.

UNGA (1988). Protection of global climate for present and future generations of mankind. United Nations General Assembly Resolution A/RES/43/53 of 6 December 1988.

UNGA (1989a). Protection of global climate for present and future generations of mankind. UN General Assembly Resolution 44/207.

UNGA (1989b). Convention on the Rights of the Child. 20 November 1989, United Nations, Treaty Series, 1577.

UNGA (1990). Protection of global climate for present and future generations of mankind. UN General Assembly Resolution 45/212.

UNGA (2000a). United Nations Millennium Declaration. Resolution adopted by the General Assembly. A/RES/55/2. New York: United Nations.

UNGA (2000b). We the Peoples: The Role of the United Nations in the Twenty-first Century. Report of the Secretary General. UN Doc. A/54/2000, 27 March 2000.

UNGA (2010a). Resolution on Human Right to Water and Sanitation (UN General Assembly Resolution A/64/292, 28 July 2010) (accessed from www.un.org/News/Press/docs/2010/ga10967.doc.htm).

UNGA (2010b). Resolution on Human Right to Water and Sanitation. Resolution A/64/292 (accessed 13 May 2013 from www.un.org/News/Press/docs/2010/ga10967.doc.htm).

UNHRC (2008). Human Rights and Climate Change. Resolution 7/23, 28 March 2008.

UNHRC (2009). Human rights and climate change. Resolution 10/4 (accessed 13 May 2013 from http://ap.ohchr.org/documents/E/HRC/resolutions/A_HRC_RES_10_4.pdf).

UNHRC (2011). Human rights and climate change. Resolution A/HRC/RES/18/22. 30 September 2011 (accessed 13 May 2013 from www.ohchr.org/Documents/Issues/ClimateChange/A.HRC.RES.18.22.pdf).

US State Department (1998). The Kyoto Protocol on Climate Change: Fact Sheet, January 15, 1998 (accessed from http://usinfo.org/enus/govern ment/forpolicy/kyoto.html).

VROM (1989). The Netherlands National Environmental Policy Plan, Ministry of Housing, Physical Planning and the Environment, The Netherlands.

VROM (1990). The Netherlands National Environmental Policy Plan Plus, Ministry of Housing, Physical Planning and the Environment, The Netherlands.

Index

activities implemented jointly (AIJ), 69, 74, 82
Ad Hoc Group on the Berlin Mandate (AGBM),
    78, 79
Ad Hoc Working Group on Further Commitments
    for Annex-I Parties under the Kyoto Protocol
    (AWG-KP), 109, 113, 124, 130, 199
Ad Hoc Working Group on Long-term Cooperative
    Action under the Convention (AWG-LCA), 112,
    124, 130, 133, 199
Ad Hoc Working Group on the Durban Platform for
    Enhanced Action, 133, 199
adaptation, 15, 22, 23, 32, 50, 65, 66, 91, 100, 109,
    115, 117, 124, 142, 165, 185, 187, 190, 200
Adaptation Fund, 106, 117, 125
Advisory Group on Greenhouse Gases (AGGG),
    43, 53
Africa, 67, 92, 96, 154, 162–4, 166
Alliance of Small Island States (AOSIS), 55, 60, 68,
    74, 96, 133, 142, 143, 153, 162, 162, 164–6
Annex/non-Annex, 63, 70, 73, 92, 96, 103,
    147–9, 171
Antigua and Barbuda, 154, 162
Argentina, 91, 96, 141, 149, 153, 162, 177
Armenia, 96, 152, 153
Asia, 92, 96
Asia-Pacific Partnership on Clean Development and
    Climate, 115, 140
assigned amounts (AAs), 82, 84, 86, 107, 126
Australia, 73, 80, 85, 86, 95, 100, 114, 115, 118, 120,
    121, 125, 128, 130, 133, 135, 152, 154, 161, 171,
    174, 175, 177, 184
Austria, 51, 115

Bali Action Plan, 101, 111, 112, 126
Bangladesh, 153, 154
BASIC, 140, 142, 154, 209
Belarus, 85, 111, 128, 149, 159, 171
Belize, 153, 174
Berlin Mandate, 62, 69
biodiversity, 26, 60, 61, 162, 201, 207

Bolivia, 130, 162
Brazil, 4, 52, 80, 114, 123, 125, 128, 140, 142, 154
Buenos Aires Plan of Action, 91
Bulgaria, 159
Bush, George H.W., 154, 193
Bush, George W., 121, 154, 183
Byrd–Hagel Resolution, 79, 94, 154

Canada, 44, 51, 85, 114, 115, 121, 128, 130, 133, 140,
    143, 149, 152, 160, 161, 171, 177, 179, 193, 199
Cancun Agreements, 130–3
capacity building, 103–4, 112, 117, 150
carbon budget, 14–17, 19, 20, 49, 56
Carbon Sequestration Leadership Forum, 114, 140
Central Asia, Caucasus and Moldova (CACAM), 96,
    120, 152
Central Group 11 (CG11), 96, 152, 159
certified emission reduction (CER), 83, 107, 126
China, 20, 51, 73, 98, 100, 114, 115, 123, 126, 128,
    140, 141, 142, 150, 153, 154, 161–2, 167, 167,
    168, 199, 210
Clean Development Mechanism (CDM), 82, 83–4,
    88, 99, 104, 106, 109, 117, 119, 139, 163, 165,
    165, 166, 168
Climate Action Network (CAN), 53, 75, 169
Climate and Clean Air Coalition to Reduce Short-
    Lived Climate Pollutants (CCAC), 142, 153
Climate Gate, 142
Climate Technology Centre and Network, 133, 135
Clinton, Bill, 77, 85, 98, 111, 121, 154
Club of Rome, 9, 9, 27
Coalition for Rainforest Nations, 109, 153
Colombia, 154, 162
Commissioner for Human Rights, 183
common but differentiated responsibilities (CBDR),
    55, 61, 64, 66, 162, 168, 173, 177–80, 190
Conference of the Parties (COPs), 61, 64, 65, 67,
    68–9, 80, 84, 91–3, 94, 96, 99, 100, 101, 103,
    104, 105, 106, 107, 109, 111, 112, 113, 116, 117,
    118, 124, 125, 126, 128, 130, 133, 134, 135, 138,

139, 140, 142, 147, 151, 160, 163, 164, 169, 184,
   185, 193, 197, 199
Copenhagen Accord, 128–30
Copenhagen Green Climate Fund, 128
corporate social responsibility, 12, 28
Costa Rica, 111, 126, 153, 154, 162
costs, 6, 29, 34, 71, 84, 173
Croatia, 111, 147, 156, 159
cyclone
   Nargis, 124
Czech Republic, 147, 159, 175

deforestation, 23, 71, 76, 80, 106, 109, 119,
   135–8, 201
Denmark, 128, 130
desertification, 61, 71, 162, 163, 201
development, 3, 10, 15, 17–19, 22–4, 94, 100, 118–19
   adaptation and, 23
   inclusive, 27
   the right to, 21, 43, 60, 182, 185, 210
   sustainable, 17, 28, 46, 50, 66, 74, 80, 83, 104, 106,
      107, 111, 118, 119, 182, 187, 189
Dominican Republic, 35, 153, 165

East Bloc, 159–60
economies in transition (EIT), 73, 96, 104, 107,
   147, 151
Ecuador, 130, 153
emission reduction units (ERUs), 83, 107
emissions
   survival versus luxury, 23, 75
   trading (ET), 82, 84, 88–90
energy
   fossil, 4, 9, 23, 29, 38, 49, 74, 75, 107, 124, 126,
      156, 161, 167
   nuclear, 29, 34, 97, 106, 124, 159, 161, 168
   renewable, 4, 29, 119, 122, 124, 187
   shale, 23, 124, 156
environmental Kuznets curve, 18, 26, 66, 94
environmentalism, 74, 120
Equatorial Guinea, 153
equitably, see justice
EU-27, 128
Europe, 86, 123, 138, 171, 185, 209
European Community, 45, 47, 51, 73
European Union, 6, 20, 78, 79, 82, 85, 86, 95, 97–8,
   115, 118, 119, 120, 122, 125, 126, 130, 134,
   135, 140, 141, 142, 143, 151, 154, 159, 177, 199,
   200, 211
   emissions trading system (EU-ETS), 156, 163
Expert Group on Technology Transfer, 105, 109, 112

fracking, 156, 167
France, 114, 121, 154
free-riding, 12, 94, 194, 200

G8, 100, 114, 123
G77, 27, 73, 96, 96, 121, 126, 142, 150, 153, 161–2,
   166, 167, 167, 211

Gabon, 153, 167
geo-engineering, 9–10, 15, 17, 22, 24, 25, 30–2, 38,
   185, 187, 190
Germany, 4, 51, 114, 115, 124, 154, 159, 177, 184,
   199, 211
Ghana, 153, 154, 162
Global Environment Facility (GEF), 65, 67, 90, 104,
   113, 115–17, 125
Global Methane Initiative, 114, 140
glocal, 3, 4, 194–5, 196
good governance, 189, 205, 205, 207
Gore, Al, 77, 78, 85, 95, 101, 111, 121, 123, 154
Green Climate Fund, 116, 130, 133, 135
green economy, 17, 24, 26–7, 114, 139, 172, 204,
   206
Greenland, 8, 130
Guyana, 153, 162

hot air, 85, 86, 95, 96
Human Development Index, 18, 27
human rights, 14, 139, 173, 174, 182–9, 190, 200,
   204, 210
Hungary, 56, 159, 159
hurricane
   Andrew, 36
   Katrina, 20, 34, 100, 142, 211
   Sandy, 20, 34, 141, 211

Iceland, 86, 95, 114, 123, 156
India, 8, 20, 52, 114, 115, 123, 128, 140, 141, 142,
   154, 162, 168–9, 171, 184
Indonesia, 128, 153, 154, 162, 167, 184
insurance, 35–7, 60, 66, 112, 126, 133, 180
Intergovernmental Panel on Climate Change (IPCC),
   5, 6, 9, 41, 44, 48, 48, 49, 53, 57, 59, 75, 77, 78,
   94, 97, 101, 104, 105, 111, 113, 120, 123, 130,
   141, 142, 169, 181, 199
   First Assessment Report, 14, 48, 53, 59, 75
   Second Assessment Report, 48, 61, 67
   Synthesis Report, 14
   Third Assessment Report, 48, 50–1, 57, 58, 66, 76,
      78, 120, 125, 173
International Council for Local Environmental
   Initiatives (ICLEI), 53, 171
International Court of Justice (ICJ), 174, 182
International Covenant on Economic, Social and
   Cultural Rights, 182, 184
International Energy Agency, 114, 156
International Partnership for Hydrogen and Fuel
   Cells in the Economy, 114, 140
IPCC, see Intergovernmental Panel on Climate
   Change
Ireland, 115, 123, 134
Italy, 114, 115, 121

Jamaica, 153, 162
Japan, 47, 60, 72, 78, 80, 85, 86, 94, 95, 98, 114, 115,
   124, 127, 133, 134, 135, 140, 143, 149, 152, 156,
   160, 179, 193, 199

joint implementation (JI), 82, 106
justice, 14, 72, 149, 163, 173

Kazakhstan, 92, 96, 128, 129, 147, 152, 171
Korea, 114, 115
Kyoto Protocol, 10, 69, 74, 78, 79, 80–93, 99, 100,
    109, 114, 119, 140, 147, 154, 156, 160, 165, 168,
    175, 176, 177, 193, 199, 200

land use, land-use change and forestry (LULUCF),
    105, 112, 165, 165
Latin America, 92, 154, 171
Latvia, 149, 159, 159
leakage, 10, 12, 13, 30, 89, 95, 111, 137, 200
least developed countries (LDCs), 92, 103, 105, 106,
    109, 121, 142, 143, 162, 166
    least developed countries fund (LDCF), 116, 117,
    118, 166
liability, 53–4, 66, 70, 72, 94, 112, 173, 180, 181, 189,
    197, 200
Liechtenstein, 128, 135, 147, 154, 156
Lithuania, 85, 149, 159, 159
litigation, 37, 173–82
lock-in, 3, 143, 156, 179

Maldives, 33, 130, 154, 164, 183, 189
Malé Declaration on Global Warming and Sea Level
    Rise, 46, 164, 183
Marrakesh Accords, 99, 101, 117
Marshall Islands, 154, 189
media, 10, 12, 21, 53, 76, 170
Mexico, 125, 126, 128, 154
Micronesia, 165, 175
migration, 11, 34, 37, 50, 66, 120, 187, 189
Millennium Development Goals, 22
mitigation, 16, 22–3, 30, 50, 66, 111, 114, 115, 142,
    165, 185, 187, 190, 200
Monaco, 135, 147, 154
Myanmar, 124

National Adaptation Programmes of Action
    (NAPAs), 103, 117, 118, 121, 135, 163,
    166
Nationally Appropriate Mitigation Actions
    (NAMAs), 111, 112, 124, 135
New Zealand, 73, 80, 85, 114, 115, 125, 133, 134,
    140, 152, 154, 161, 193, 199
Nigeria, 153, 162, 183
no-harm principle, 54, 66, 190, 197
Nobel Peace Prize, 10, 101, 111, 123
Non-aligned Movement, 45, 58
non-governmental organizations (NGOs), 53, 96–7,
    107, 120, 121, 169, 170, 177
Noordwijk Conference on Climate Change, 51
Noordwijk Declaration on Climate Change, 46,
    55, 56
no-regrets strategy, 33, 57, 62
North–South, 4, 14–17, 26, 49, 65, 68, 69, 73, 74, 142,
    149, 154, 156, 172, 195, 203

Norway, 80, 86, 95, 114, 115, 126, 127, 134, 152,
    154, 156

Obama, Barack, 123, 126, 142, 156
Office of the High Commissioner for Human Rights
    (OHCHR), 184
official development assistance (ODA), 57, 106,
    126, 189
Organization for Economic Co-operation and
    Development (OECD), 119, 149
Organization of the Petroleum Exporting Countries
    (OPEC), 74, 96, 115, 121, 153, 162, 162, 166, 167

Pakistan, 124, 149, 153, 162
Papua New Guinea, 111, 153, 165
Peru, 154, 174
Poland, 56, 159, 159
polluter pays principle, 61, 66, 126, 180, 190, 197
Portugal, 86, 123
precautionary principle, 6, 66, 162, 174, 180, 181,
    190, 197, 206
preparedness, 50, 57, 117, 133
prevention, 34, 54, 133, 137, 185

Qatar, 149, 162
Quantified Emission Limitation and Reduction
    Objective (QELRO), 69, 82

recession, 138–9, 140, 143
REDD+, 111, 203
Reducing Emissions from Deforestation (RED),
    111, 112
Reducing Emissions from Deforestation and Forest
    Degradation (REDD), 111, 112, 124, 126, 133,
    135–8, 139, 142, 154, 165, 166, 187, 199, 200
resilience, 23, 27, 32, 33, 37, 100, 142
respective capabilities, 61, 64, 66, 169, 173
Rio Declaration, *see* United Nations, Declaration on
    Environment and Development
Rio+20 conference, 193
Romania, 93, 159, 159
rule of law, 203–6, 207, 208, 209, 211
Russia, 8, 85, 86, 95, 96, 98, 100, 105, 114, 133, 140,
    143, 156, 159, 160–1, 179, 193, 199

Samoa, 153, 154
Saudi Arabia, 52, 109, 167
scepticism, 5–11, 12, 21, 49, 75, 94, 120, 123, 140,
    142, 155, 194, 196, 203, 203
Singapore, 130, 149, 165
Slovakia, 147, 159
Slovenia, 147, 159, 159
Small Island Developing States (SIDS), 164, 166
South Africa, 114, 123, 126, 140, 142, 154, 162,
    171
South Korea, 128, 130, 140, 154
Soviet Union, 47, 51
Spain, 115, 123, 154
Stern report, 100–1, 111

subsidiary body for implementation (SBI), 65, 91, 104, 112, 133
subsidiary body for scientific and technological advice (SBSTA), 65, 91, 103, 104, 105, 109, 112
Switzerland, 115, 126, 128, 134, 135, 152, 154

Tanzania, 52, 154, 162
technocratic, 4, 49, 197
technology, 15, 57, 104–5, 117, 124, 150, 199
Thailand, 153, 154
The Hague Declaration, 45, 139
The Netherlands, 33, 51, 115, 121, 154, 171, 177
Tokyo Conference on the Global Environment, 45, 55
Toronto Declaration, 56, 85
Turkey, 56, 66, 73, 92, 103, 149, 171, 193
Turkmenistan, 96, 152
Tuvalu, 165, 189

Ukraine, 85, 86, 96, 159, 160
umbrella group, 152, 160, 160
UNDP, *see* United Nations, Development Programme
United Kingdom, 98, 114, 115, 134, 154, 159, 177, 184, 199
United Nations, 133, 205
    Charter, 19, 46, 54, 184, 209, 211
    Conference on Environment and Development (UNCED), 26, 48, 59, 60–1, 100
    Declaration on Environment and Development, 61, 62, 180, 186, 209
    Development Programme (UNDP), 60, 115, 117, 118, 139, 140, 141, 184

Environment Programme (UNEP), 44, 48, 58, 60, 65, 115, 119, 135, 139, 183
Framework Convention on Climate Change, 20, 26, 60, 62–8, 75, 115, 133, 139, 140, 154, 165, 173, 179, 185, 193–4, 195, 197, 199, 202, 208
General Assembly (UNGA), 19, 37, 41, 44, 46, 48, 54, 58, 166, 174, 184
Human Rights Council, 140, 184
Intergovernmental Negotiating Committee (INC), 41, 48, 59–61, 67, 68–9
Security Council, 101, 123, 167
United States of America, 6, 8, 20, 44, 47, 51, 56, 60, 61, 66, 69, 72, 73, 77, 78, 79, 80, 84, 85, 86, 91, 93, 95, 96, 97–8, 99, 100, 101, 109, 114, 115, 118, 119, 120, 121, 123, 124, 125, 126, 128, 130, 133, 140, 141, 142, 143, 149, 151, 154–6, 159, 160, 161, 167, 168, 168, 171, 172, 174, 175, 177, 179, 183, 188, 193, 199, 200, 209, 210
Uruguay, 153, 154
USA, *see* United States of America
USAID, 114
Uzbekistan, 96, 152

Vanuatu, 153, 164, 189
vulnerability, 23, 32, 33, 65, 97, 100, 104, 105, 107, 109, 153, 166, 203

World Bank, 60, 65, 113, 115, 119, 126, 139, 209
World Climate Conference, 41, 47, 48, 50, 54, 55
World Summit on Sustainable Development, 100, 107
World Trade Organization (WTO), 160, 175